# Steam Tables in SI-Units

# Wasserdampftafeln

Concise Steam Tables in SI-Units (Student's Tables)
Properties of Ordinary Water Substance
up to 1000 °C and 100 Megapascal

Kurzgefaßte Dampftafeln in SI-Einheiten (Studententafeln)
Zustandsgrößen von gewöhnlichem Wasser und Dampf
bis 1000 °C und 1000 bar

Third, Enlarged Edition

Edited by

Ulrich Grigull · Johannes Straub
Peter Schiebener

Springer-Verlag Berlin Heidelberg GmbH 1990

o. Professor em. Dr.-Ing., Dr.-Ing. E. h. Ulrich Grigull
a. o. Professor Dr.-Ing. habil. Johannes Straub
Dr.-Ing. Peter Schiebener

Lehrstuhl A für Thermodynamik, Technische Universität München,
Arcisstraße 21, D-8000 München 2

**Additional material to this book can be downloaded from http://extras.springer.com**

CIP-Titelaufnahme der Deutschen Bibliothek
*Grigull, Ulrich:*
Steam tables in SI units : concise steam tables in SI-units
(student's tables) ; properties of ordinary water substance up to
1000 °C and 100 megapascal = Wasserdampftafeln / ed. by
Ulrich Grigull ; Johannes Straub ; Peter Schiebener. — 3., enl.
ed. — Berlin ; Heidelberg ; New York ; Tokyo ; Hong Kong :
Springer, 1990

ISBN 978-3-540-51888-4      ISBN 978-3-642-95604-1 (eBook)
DOI 10.1007/978-3-642-95604-1

NE: Straub, Johannes:; Schiebener, Peter:; HST

2160/3020-543210 — Gedruckt auf säurefreiem Papier

# Foreword to the Third Edition

The tables and diagrams concerning the properties of ordinary water substance – as offered in this booklet – are mainly meant for use by students at universities and colleges so that they may be able to solve problems in the fields of power and chemical engineering, where water and steam are serving as working or process medium.

On the other hand the tables and diagrams should support engineers in research work and industrial practice to obtain a quick and reliable general view of the properties of water substance.

All tabulated values given in the present Third Edition of the "Student's Tables" were recalculated. The thermodynamic properties have been calculated according to the 1984 IAPS Formulation for the Thermodynamic Properties of Ordinary Water Substance for Scientific and General Use, employing reduced iteration limits as compared to earlier calculations. The cubic expansion coefficient was added. The remaining properties were calculated according to the current releases of the International Association for the Properties of Steam (IAPS). The increments for temperature and pressure for the saturation tables were decreased. In addition, ten properties were added. Saturation pressure tables were also added to complement the existing saturation temperature tables.

The table of the properties at 0.1 Megapascal was expanded to a temperature of − 30 °C. The three new $h, s$-diagrams for compressed water will be useful (amongst others) in geophysical and in jet cutting applications. A new table shows the thermodynamic relationships between the properties and their derivatives. These identities are valid for any substance.

The editors are hopeful that the "Student's Tables" in this expanded edition are again well received.

Munich, September 1989                                            The Editors

# Vorwort zur dritten Auflage

Die hier vorgelegten Tafeln und Diagramme über die Eigenschaften von gewöhnlichem Wasser sind in erster Linie für den Gebrauch der Studenten an Universitäten und Fachhochschulen bestimmt. Diese sollen damit Probleme aus der Energietechnik und der Verfahrenstechnik lösen können, bei denen Wasser und Wasserdampf als Arbeits- oder Prozeßmedium eine Rolle spielen.

Die Tafeln und Diagramme sollen aber auch dem Ingenieur in Forschung und Betrieb einen schnellen und zuverlässigen Überblick über die Eigenschaften der Substanz Wasser geben.

Für diese dritte Auflage der „Studententafeln" wurden sämtliche Tafelwerte neu berechnet. Die thermodynamischen Zustandsgrößen wurden nach der 1984 IAPS Formulation for the Thermodynamic Properties of Ordinary Water Substance for Scientific and General Use berechnet, und zwar mit einer verringerten Iterationsschranke gegenüber früheren Rechnungen. Der Volumenausdehnungskoeffizient wurde neu aufgenommen. Die übrigen Größen wurden nach den heute gültigen „Releases" der International Association for the Properties of Steam (IAPS) berechnet. In den Tafeln für die Sättigungsgrößen wurde die Stufung für Temperatur und Druck wesentlich verfeinert. Ferner wurden zehn Zustandsgrößen neu aufgenommen. Neben den bisherigen Temperaturtafeln wurden auch Drucktafeln für die Sättigungswerte hinzugefügt.

Die Tafel für die Zustandsgrößen bei 1 bar wurde bis − 30 °C erweitert. Drei neue $h, s$-Diagramme für Druckwasser sollen Anwendungen z.B. in der Geophysik und beim Strahlschneiden (jet cutting) dienen. Eine neue Tabelle zeigt thermodynamische Beziehungen zwischen den Zustandsgrößen und ihren Ableitungen. Diese Identitäten gelten für beliebige Substanzen.

Die Herausgeber wünschen den Studententafeln auch in dieser erweiterten Ausgabe eine gute Aufnahme.

München, im September 1989                                      Die Herausgeber

# Contents

foldout
tables

# Inhaltsverzeichnis

Ausklapp-
tafeln

# I. Quantities, Constants, Units

# I. Größen, Konstanten, Einheiten

## Table A. Quantities and Units Presented
## Tabelle A. Dargestellte Größen und Einheiten

| Quantity, Größe | | SI unit, SI-Einheit | |
|---|---|---|---|
| Name | Symbol | Name | Symbol |
| pressure<br>Druck | $p$ | pascal<br>Pascal | Pa |
| temperature<br>Temperatur | $T$ | kelvin<br>Kelvin | K |
| Celsius temperature<br>Celsius-Temperatur | $t$ | degree Celsius<br>Grad Celsius | °C |
| specific volume<br>spezifisches Volumen | $v$ | cubic metre per kilogram<br>Kubikmeter durch Kilogram | $\dfrac{m^3}{kg}$ |
| density<br>Dichte | $\rho$ | kilogram per cubic metre<br>Kilogramm durch Kubikmeter | $\dfrac{kg}{m^3}$ |
| specific internal energy<br>spezifische innere Energie | $u$ | joule per kilogram<br>Joule durch Kilogramm | $\dfrac{J}{kg}$ |
| specific entropy<br>spezifische Entrope | $s$ | joule per kilogram kelvin<br>Joule durch Kilogrammkelvin | $\dfrac{J}{kg\,K}$ |
| specific enthalpy<br>spezifische Enthalpie | $h$ | joule per kilogram<br>Joule durch Kilogramm | $\dfrac{J}{kg}$ |
| specific heat capacity<br>at constant pressure<br>spezifische Wärmekapazität<br>bei konstantem Druck | $c_p$ | joule per kilogram kelvin<br><br>Joule durch Kilogrammkelvin | $\dfrac{J}{kg\,K}$ |
| specific heat capacity<br>at constant volume<br>spezifische Wärmekapazität<br>bei konstantem Volumen | $c_v$ | joule per kilogram kelvin<br><br>Joule durch Kilogrammkelvin | $\dfrac{J}{kg\,K}$ |
| specific heat capacity<br>at saturation<br>spezifische Wärmekapazität<br>bei Sättigung | $c_{sat}$ | joule per kilogram kelvin<br><br>Joule durch Kilogrammkelvin | $\dfrac{J}{kg\,K}$ |
| cubic expansion coefficient<br>Volumenausdehnungskoeffizient | $\alpha_p$ | reciprocal kelvin<br>reziprokes Kelvin | $\dfrac{1}{K}$ |
| isochoric pressure coefficient<br>isochorer Spannungskoeffizient | $\beta_v$ | reciprocal kelvin<br>reziprokes Kelvin | $\dfrac{1}{K}$ |
| isentropic pressure coefficient<br>isentroper Spannungskoeffizient | $\beta_s$ | reciprocal kelvin<br>reziprokes Kelvin | $\dfrac{1}{K}$ |
| isenthalpic Joule-Thomson<br>coefficient<br>isenthalper Drosselkoeffizient | $\delta_h$ | kelvin per pascal<br><br>Kelvin durch Pascal | $\dfrac{K}{Pa}$ |
| isothermal Joule-Thomson<br>coefficient<br>isothermer Drosselkoeffizient | $\delta_T$ | cubic metre per kilogram<br><br>Kubikmeter durch Kilogramm | $\dfrac{m^3}{kg}$ |

| Units Employed / Verwendete Einheiten | Remarks / Bemerkungen | Calculated from / Berechnet nach |
|---|---|---|
| $MPa = 10^6\ Pa$ <br> $bar = 10^5\ Pa$ | $Pa = \dfrac{N}{m^2} = \dfrac{J}{m^3} = \dfrac{kg}{s^2\,m}$ | – |
| $K$ | $T$ is defined by the IPTS-68 | – |
| $°C$ | $t = T - T_0$ <br> $T_0 = 273.15\ K$ | – |
| $\dfrac{dm^3}{kg} = 10^{-3}\ \dfrac{m^3}{kg}$ | | [1] |
| $\dfrac{kg}{m^3}$ | $\rho = \dfrac{1}{v}$ | [1, 6] |
| $\dfrac{kJ}{kg} = 10^3\ \dfrac{J}{kg}$ | | [1] |
| $\dfrac{kJ}{kg\ K} = 10^3\ \dfrac{J}{kg\ K}$ | | [1] |
| $\dfrac{kJ}{kg} = 10^3\ \dfrac{J}{kg}$ | $h = u + p\,v$ | [1] |
| $\dfrac{kJ}{kg\ K} = 10^3\ \dfrac{J}{kg\ K}$ | $c_p = \left(\dfrac{\partial h}{\partial T}\right)_p$ | [1, 6] |
| $\dfrac{kJ}{kg\ K} = 10^3\ \dfrac{J}{kg\ K}$ | $c_v = \left(\dfrac{\partial u}{\partial T}\right)_v$ | [1] |
| $\dfrac{kJ}{kg\ K} = 10^3\ \dfrac{J}{kg\ K}$ | $c_{sat} = T\left(\dfrac{\partial s}{\partial T}\right)_{sat}$ | [1] |
| $\dfrac{1}{kK} = \dfrac{10^{-3}}{K}$ | $\alpha_p = \dfrac{1}{v}\left(\dfrac{\partial v}{\partial T}\right)_p$ | [1] |
| $\dfrac{1}{K}$ | $\beta_v = \dfrac{1}{p}\left(\dfrac{\partial p}{\partial T}\right)_v$ | [1] |
| $\dfrac{1}{K}$ | $\beta_s = \dfrac{1}{p}\left(\dfrac{\partial p}{\partial T}\right)_s$ | [1] |
| $\dfrac{K}{MPa} = \dfrac{10^{-6}\ K}{Pa}$ | $\delta_h = \left(\dfrac{\partial T}{\partial p}\right)_h$ | [1] |
| $\dfrac{dm^3}{kg} = 10^{-3}\ \dfrac{m^3}{kg}$ | $\delta_T = \left(\dfrac{\partial h}{\partial p}\right)_T$ | [1] |

3

**Table A. Continued**
**Tabelle A. Fortsetzung**

| Quantity, Größe | | SI unit, SI-Einheit | |
| --- | --- | --- | --- |
| Name | Symbol | Name | Symbol |
| isothermal compressibility<br>isotherme Kompressibilität | $\chi_T$ | reciprocal pascal<br>reziprokes Pascal | $\dfrac{1}{Pa}$ |
| velocity of sound<br>Schallgeschwindigkeit | $c$ | metre per second<br>Meter durch Sekunde | $\dfrac{m}{s}$ |
| thermal conductivity<br>Wärmeleitfähigkeit | $\lambda$ | watt per kelvin metre<br>Watt durch Kelvinmeter | $\dfrac{W}{K\,m}$ |
| thermal diffusivity<br>Temperaturleitfähigkeit | $a$ | square metre per second<br>Quadratmeter durch Sekunde | $\dfrac{m^2}{s}$ |
| viscosity<br>Viskosität | $\eta$ | pascal second<br>Pascalsekunde | Pa s |
| kinematic viscosity<br>kinematische Viskosität | $\nu$ | square metre per second<br>Quadratmeter durch Sekunde | $\dfrac{m^2}{s}$ |
| Prandtl number<br>Prandtl-Zahl | $Pr$ | one<br>Eins | 1 |
| surface tension<br>Oberflächenspannung | $\sigma$ | newton per metre<br>Newton durch Meter | $\dfrac{N}{m}$ |
| Laplace coefficient<br>Laplace-Koeffizient | $b$ | metre<br>Meter | m |

**References for Table A**
**Literatur zu Tabelle A**

[1] Release on The IAPS Formulation 1984 for the Thermodynamic Properties of Ordinary Water Substance for Scientific and General Use. December 1984. Available by Dr. Howard J. White, Jr. Standard Reference Data. National Institute of Standards and Technology. Gaithersburg, MD 20 899, USA.

[2] Release on The IAPS Formulation 1985 for the Thermal Conductivity of Ordinary Water Substance. November 1985. Available as [1].

[3] Release on The IAPS Formulation 1985 for the Viscosity of Ordinary Water Substance. November 1985. Available as [1].

[4] Release on Surface Tension of Water Substance. January 1976. Available as [1].

[5] Saul, A.: Eine Fundamentalgleichung für den fluiden Zustandsbereich von Wasser bis zu Drücken von 25 000 MPa und Temperaturen von 1273 K. Fortschritt-Berichte VDI, Reihe 3, Nr. 149. Düsseldorf: VDI-Verlag 1988. – Saul, A.; Wagner, W.: A Fundamental Equation for Water covering the Range from the Melting Line to 1273 K at Pressures up to 25 000 MPa. J. Phys. Chem. Ref. Data (to be published).

[6] Sato, H.: An Equation of State for the Thermodynamic Properties of Water in the Liquid Phase including the Metastable State. Proc. 11th Int. Conf. Properties of Steam, Prague 1989 (to be published).

[7] Release on the Pressure along the Melting and Sublimation Curve of Ordinary Water (to be published).

| Units Employed / Verwendete Einheiten | Remarks / Bemerkungen | Calculated from / Berechnet nach |
|---|---|---|
| $\dfrac{1}{\text{GPa}} = \dfrac{10^{-9}}{\text{Pa}}$ | $\chi_T = -\dfrac{1}{v}\left(\dfrac{\partial v}{\partial p}\right)_T$ | [1] |
| $\dfrac{\text{km}}{\text{s}} = 10^3\,\dfrac{\text{m}}{\text{s}}$ | $c = \left(\dfrac{\partial p}{\partial \rho}\right)_s^{1/2}$ | [1] |
| $\dfrac{\text{mW}}{\text{K m}} = 10^{-3}\,\dfrac{\text{W}}{\text{K m}}$ | | [2, 6] |
| $10^{-6}\,\dfrac{\text{m}^2}{\text{s}}$ | $a = \dfrac{\lambda}{\rho\,c_p}$ | [1, 2, 6] |
| $\mu\text{Pa s} = 10^{-6}\,\text{Pa s}$ | $\text{Pa s} = \dfrac{\text{N s}}{\text{m}^2} = \dfrac{\text{kg}}{\text{s m}}$ | [3, 6] |
| $10^{-6}\,\dfrac{\text{m}^2}{\text{s}}$ | $\nu = \dfrac{\eta}{\rho}$ | [1, 3, 6] |
| $1$ | $Pr = \dfrac{\eta\,c_p}{\lambda} = \dfrac{\nu}{a}$ | [1, 2, 3, 6] |
| $\dfrac{\text{mN}}{\text{m}} = 10^{-3}\,\dfrac{\text{N}}{\text{m}}$ | $\dfrac{\text{mN}}{\text{m}} = \dfrac{\text{dyn}}{\text{cm}}$ | [4] |
| $\text{mm} = 10^{-3}\,\text{m}$ | $b = \left(\dfrac{\sigma/g_n}{\rho_{\text{liq}} - \rho_{\text{vap}}}\right)^{1/2}$ $g_n = 9.80665\ \text{m/s}^2$ | [1, 4] |

state of saturation      Sättigungszustand

for water index '      für Wasser Index '
for steam index "      für Dampf Index "

$\Delta$ difference of the values at saturation, e.g. $\Delta h = h'' - h' = r$ (specific enthalpy of evaporation)
$\Delta$ Differenz der Sättigungswerte, z.B. $\Delta h = h'' - h' = r$ (spezifische Verdampfungsenthalpie)

## Table B. Constant Quantities of Ordinary Water

## Tabelle B. Konstante Größen von gewöhnlichem Wasser

| | |
|---|---|
| molar (universal) gas constant[a]<br>molare (allgemeine) Gaskonstante[a] | $R = (8.314510 \pm 0.00007)$ J/mol K |
| molar mass[b]<br>molare Masse[b] | $M_i (H_2O) = 18.0152$ kg/kmol |
| specific gas constant<br>spezifische Gaskonstante | $R_i (H_2O) = 0.461527$ kJ/kg K<br>$R_i (H_2O) = R/M_i (H_2O)$ |
| critical temperature[c]<br>kritische Temperatur[c] | $T_c = (647.14 + \delta)$ K; $t_c = (373.99 + \delta)\,^\circ$C; $-0.10 \leqslant \delta \leqslant 0.10$ |
| critical pressure[c]<br>Kritischer Druck[c] | $p_c = (22.064 + 0.27\,\delta \pm 0.005)$ MPa |
| critical density[c]<br>kritische Dichte | $\rho_c = (322 \pm 3)$ kg/m$^3$ |
| critical specific volume<br>kritisches spezifisches Volumen | $v_c = (3.11 \pm 0.03)$ dm$^3$/kg |

[a]  Codata Newsletter October 1986. − PTB Mitteilungen 97 (1987) 498−507.
[b]  Pure and Applied Chemistry 56 (1984) 653−674.
[c]  Release on IAPS Statement 1983 of Values of Temperature, Pressure and Density of Ordinary and Heavy Water Substance at their Respective Critical Points. May 1983.

The numerical values of table B are the present best values. They do not necessarily coincide with the constants in the formulations.

The internal energy and the entropy of liquid water at the triple point are zero by definition (resolution of the 5th International Conference on the Properties of Steam, London 1956).

Die Zahlenwerte der Tabelle B sind die derzeitigen Bestwerte. Sie müssen nicht mit den Konstanten der Formulationen übereinstimmen.

Die innere Energie und die Entropie von flüssigem Wasser beim Tripelpunkt sind nach Definition Null (Beschluß der 5th International Conference on the Properties of Steam, London 1956).

## Table C. Conversion of Units

## Tabelle C. Umrechnung von Einheiten

Where the last digit appears in bold face the numerical value is exact.

Die halbfett gesetzte letzte Ziffer bedeutet, daß der zugehörige Zahlenwert genau ist.

### C 1. Pressure

### C 1. Druck

$$Pa = N/m^2 = J/m^3 = kg/s^2\ m$$

| | | |
|---|---|---|
| bar | 1 bar | $= 10^5$ Pa $= 0.1$ MPa |
| standard atmosphere | 1 atm | $= 101325$ Pa |
| Torr | 1 Torr | $= 1$ atm$/760 = 133.322$ Pa |
| kilogram-force[a] per square metre | 1 kgf/m$^2$ | $= 9.80665$ Pa |
| technical atmosphere | 1 at | $= 1$ kgf/cm$^2 = 0.967841$ atm $= 98066.5$ Pa |
| conventional millimetre of water | 1 mmH$_2$O | $= 10^{-4}$ at $= 9.80665$ Pa |
| conventional millimetre of mercury | 1 mmHg | $= 13.595$ mmH$_2$O $= 133.322$ Pa |
| pound force per square inch | 1 lbf/in$^2$ | $= 6894.76$ Pa |
| inch of water | 1 inH$_2$O | $= 249.089$ Pa |
| inch of mercury | 1 inHg | $= 3386.38$ Pa |

[a] The unit "kilogram-force (kgf)" is also called "kilopond (kp)".

[a] Die Einheit „kilogram-force (kgf)" wird auch „Kilopond (kp)" genannt.

### C 2. Temperature

### C 2. Temperatur

Between the numerical values                    Zwischen den Zahlenwerten

$$T_K,\ T_R,\ t_C,\ t_F$$

of a temperature in the Kelvin-, Rankine-, Celsius-, Fahrenheitscale, the following relations exist:

einer Temperatur in der Kelvin-, Rankine, Celsius- und Fahrenheit-Skala bestehen folgende Beziehungen:

$$T_K = 273.15 + t_C = (5/9)\ T_R$$
$$T_R = 459.67 + t_F = 1.8\ T_K$$
$$t_C = (5/6)\ (t_F - 32) = T_K - 273.15$$
$$t_F = 1.8\ t_C + 32 = T_R - 459.67$$

## C 3. Energy, Work, Quantity of Heat

## C 3. Energie, Arbeit, Wärmemenge

$$J = Nm = Pa\ m^3 = Ws = kg\ m^2/s^2$$

| | | |
|---|---|---|
| erg | 1 erg $= 1$ dyn cm | $= 10^{-7}$ J |
| kilowatt hour | 1 kWh $= 3.6 \cdot 10^6$ J | $= 3.6$ MJ |
| kilogram-force metre | 1 kgf m $= 9.80665$ J | |
| 15° calorie | 1 cal$_{15}$ $= 4.1855$ J | |
| thermochemical calorie | 1 cal$_{th}$ $= 4.184$ J | |
| International Table calorie[a] | 1 cal$_{IT}$ $= 4.1868$ J | |
| International Table British thermal unit[a] | 1 Btu$_{IT}$ $= 778.169$ ft lbf $= 1055.06$ J | |
| foot pound force | 1 ft lbf $= 1.35582$ J | |

[a] The 5th International Conference on the Properties of Steam, London 1956, adopted the "International Table calorie" and defined the "International Table British thermal unit" by 1 Btu$_{IT}$/lb $= 2.326$ J/g (exactly). This is equivalent to 1 kcal$_{IT}$/kg $= 1.8$ Btu$_{IT}$/lb (exactly).

[a] Die 5th International Conference on the Properties of Steam, London 1956, nahm die „International Table calorie" an und definierte die „International Table British thermal unit" durch die Gleichung 1 Btu$_{IT}$ $= 2.326$ J/g (genau). Damit ist auch 1 kcal$_{IT}$/kg $= 1.8$ Btu$_{IT}$/lb (genau).

## C 4. Specific Heat Capacity, Specific Entropy, Specific Gas Constant

## C 4. Spezifische Wärmekapazität, Spezifische Entropie, Spezifische Gaskonstante

| | | |
|---|---|---|
| IT calorie per gram kelvin | 1 cal$_{IT}$/g K | $= 4186.8$ J/kg K |
| thermochemical calorie per gram kelvin | 1 cal$_{th}$/g K | $= 4184$ J/kg K |
| IT British thermal unit per pound degree Rankine[a] | 1 Btu$_{IT}$/lb °R | $= 4186.8$ J/kg K |

[a] In consequence of the definition of the Btu$_{IT}$ (compare table C 3)
1 Btu$_{IT}$/lb °R $= 1$ kcal$_{IT}$/kg K (exactly).

[a] Wegen der Definition der Btu$_{IT}$ (vgl. Tabelle C 3) gilt auch
1 Btu$_{IT}$/lb °R $= 1$ kcal$_{IT}$/kg K (genau).

## C 5. Thermal Conductivity

## C 5. Wärmeleitfähigkeit

| | | |
|---|---|---|
| IT calorie per second centimetre kelvin | 1 cal$_{IT}$/s cm K | $= 418.68$ W/K m |
| IT kilocalorie per hour metre kelvin | 1 kcal$_{IT}$/h m K | $= 1.163$ W/K m |
| thermochemical calorie per second centimetre kelvin | 1 cal$_{th}$/s cm K | $= 418.4$ W/K m |
| IT British thermal unit per second foot degree Rankine | 1 Btu$_{IT}$/s ft °R | $= 6230.67$ W/K m |
| IT British thermal unit per hour foot degree Rankine | 1 Btu$_{IT}$/h ft °R | $= 1.73074$ W/K m |

## C 6. Viscosity

## C 6. Viskosität

$$Pa\ s = N\ s/m^2 = kg/s\ m$$

| | | |
|---|---|---|
| poise | 1 P $= 1$ dyn s/cm$^2$ $= 1$ g/s cm $= 0.1$ Pa s | |
| kilogram-force second per square metre | 1 kgf s/m$^2$ $= 9.80665$ Pa s | |
| pound-force second per square foot | 1 lbf s/ft$^2$ $= 47.8802$ Pa s | |
| pound per foot second | 1 lb/ft s $= 1.48816$ Pa s | |

## C 7. Thermal Diffusivity, Kinematic Viscosity, Diffusion Coefficient
## C 7. Temperaturleitfähigkeit, Kinematische Viskosität, Diffusionskoeffizient

| | |
|---|---|
| square metre per hour | $1 \text{ m}^2/\text{h} = 2.\overline{7} \cdot 10^{-4} \text{ m}^2/\text{s}$ |
| square foot per second | $1 \text{ ft}^2/\text{s} = 92.90304 \cdot 10^{-3} \text{ m}^2/\text{s}$ |
| square foot per hour | $1 \text{ ft}^2/\text{h} = 25.8064 \cdot 10^{-6} \text{ m}^2/\text{s}$ |

The unit of the kinematic viscosity $\text{cm}^2/\text{s}$ is also called "stokes" ($1 \text{ St} = 1 \text{ cm}^2/\text{s} = 10^{-4} \text{ m}^2/\text{s}$).

Die Einheit der kinematischen Viskosität $\text{cm}^2/\text{s}$ wird auch „Stokes" genannt ($1 \text{ St} = 1 \text{ cm}^2/\text{s} = 10^{-4} \text{ m}^2/\text{s}$).

## C 8. Surface Tension
## C 8. Oberflächenspannung

$\text{N/m} = \text{J/m}^2 = \text{kg/s}^2$

| | |
|---|---|
| dyne per centimetre | $1 \text{ dyn/cm} = 1 \text{ erg/cm}^2 = 10^{-3} \text{ N/m}$ |
| kilogram-force per metre | $1 \text{ kgf/m} = 9.80665 \text{ N/m}$ |

**Table D. Thermodynamic Relationships between the Properties $p, v, T, s, u, h$ and their Derivatives[a]**

**Tabelle D. Thermodynamische Beziehungen zwischen den einfachen Zustandsgrößen $p, v, T, s, u, h$ und ihren Ableitungen[a]**

$$\delta_h = \frac{v}{c_p}(\alpha_p T - 1) ; \quad \delta_T = -c_p \, \delta_h \tag{1}$$

$$p \, \beta_v = \frac{\alpha_p}{\chi_T} ; \quad p \, \beta_s = \frac{c_p}{v \, T \, \alpha_p} = \frac{1}{\delta_h + v/c_p} \tag{2}$$

$$c^2 = \frac{c_p \, v}{c_v \, \chi_T} \tag{3}$$

$$\left.\frac{dp}{dT}\right|_{sat} = \frac{\Delta s}{\Delta v} = \frac{r}{T \, \Delta v} \quad \text{Clausius-Clapeyron equation} \tag{4}$$

$$\frac{dr}{dT} = \Delta c_p + \Delta \delta_T \left.\frac{dp}{dT}\right|_{sat} \tag{5}$$

$$\frac{dr}{dT} = \Delta c_p + \frac{r}{T} - \frac{r \, \Delta(v \, \alpha_p)}{\Delta v} \quad \text{Clausius-Planck equation} \tag{6}$$

$$c_{sat} = T \left(\frac{\partial s}{\partial T}\right)_{sat} = c_p - v \, T \, \alpha_p \left.\frac{dp}{dT}\right|_{sat} \tag{7a}$$

$$c_{sat} = c_p + (\delta_T - v) \left.\frac{dp}{dT}\right|_{sat} \tag{7b}$$

$$\Delta c_{sat} = \Delta c_p - \frac{r}{T} + \Delta \delta_T \left.\frac{dp}{dT}\right|_{sat} = \frac{dr}{dT} - \frac{r}{T} \tag{8}$$

The following limiting values are valid for the critical point:

Für den kritischen Punkt gelten folgende Grenzwerte:

$$\left.\frac{\mathrm{d}p}{\mathrm{d}T}\right|_{\mathrm{sat}} = \frac{1}{\delta_{\mathrm{h}}} = p\,\beta_{\mathrm{v}} = p\,\beta_{\mathrm{s}} \tag{9}$$

$$\left.\frac{\mathrm{d}p}{\mathrm{d}T}\right|_{\mathrm{sat}} = 0.2679\,\frac{\mathrm{MPa}}{\mathrm{K}} \quad \text{for ordinary water substance} \tag{9a}$$

$$\delta_{\mathrm{h}} = -\frac{\delta_{\mathrm{T}}}{c_{\mathrm{p}}} = \frac{\chi_{\mathrm{T}}}{\alpha_{\mathrm{p}}} = \frac{v\,T\,\alpha_{\mathrm{p}}}{c_{\mathrm{p}}} = -\frac{v\,T\,\chi_{\mathrm{T}}}{\delta_{\mathrm{T}}} \tag{10}$$

$$\delta_{\mathrm{h}} = 3.733\,\frac{\mathrm{K}}{\mathrm{MPa}} \quad \text{for ordinary water substance} \tag{10a}$$

$$c^2\,c_{\mathrm{v}} = \frac{v\,c_{\mathrm{p}}}{\chi_{\mathrm{T}}} = \frac{T}{(\rho\,\delta_{\mathrm{h}})^2} \tag{11}$$

$$c^2\,c_{\mathrm{v}} = 447.9\,\frac{\mathrm{kJ}^2}{\mathrm{kg}^2\,\mathrm{K}} \quad \text{for ordinary water substance} \tag{11a}$$

The values of the properties $\alpha_{\mathrm{p}}$, $c_{\mathrm{p}}$, $\delta_{\mathrm{T}}$ and $\chi_{\mathrm{T}}$ approach infinity in the same order at the critical point.

Die Größen $\alpha_{\mathrm{p}}$, $c_{\mathrm{p}}$, $\delta_{\mathrm{T}}$ und $\chi_{\mathrm{T}}$ erreichen im kritischen Punkt den Wert unendlich in der gleichen Ordnung.

[a] See also: Grigull, U.; Bach, J.; Reimann, M.: The Properties of Water and Steam according to "The 1968 IFC Formulation" (in German). Wärme- und Stoffübertragung 1 (1968) 202-213.

# II. Tables of the Properties of Ordinary Water Substance

# II. Tafeln der Zustandsgrößen von gewöhnlichem Wasser und Wasserdampf

**Table 1.** Thermodynamic and Transport Properties of Liquid Water and Superheated Steam inclusive the Saturation Values from the Triple Point to 1000 °C and 100 MPa: $p, t, v, h, u, s, c_p, \alpha_p, \lambda, \eta$ (Pressure Table).
The values at 0 °C in the tables for $p \leqslant 0.1$ MPa relate to subcooled water.

**Tafel 1.** Thermodynamische und Transportgrößen von Wasser und überhitztem Dampf einschließlich der Sättigungswerte vom Tripelpunkt bis 1000 °C und 1000 bar: $p, t, v, h, u, s, c_p, \alpha_p, \lambda, \eta$ (Drucktafel).
Die Werte bei 0 °C in den Tafeln für $p \leqslant 1$ bar gelten für unterkühltes Wasser.

Table 1

$$p = 0.0006117\,\text{MPa} = 0.006117\ \text{bar}$$

| $t$ | $v$ | $h$ | $u$ | $s$ | $c_p$ | $\alpha_p$ | $\lambda$ | $\eta$ |
|---|---|---|---|---|---|---|---|---|
| °C | dm³/kg | kJ/kg | kJ/kg | kJ/kg K | kJ/kg K | $10^{-3}$/K | mW/K m | µPa s |
| 0.0 | 1.0002 | −0.04 | −0.04 | −0.0002 | 4.229 | −0.081 | 561.0 | 1792 |
| 0.01 | triple point | | | | | | | |
| solid | 1.0908 | −333.4 | −333.4 | −1.221 | 1.93 | 0.1 | 2.2 | — |
| liquid | 1.0002 | 0.0 | 0 | 0 | 4.229 | −0.080 | 561.0 | 1791 |
| vapour | 205986 | 2500 | 2374 | 9.154 | 1.868 | 3.672 | 17.07 | 9.22 |
| 5.0 | 209913 | 2509 | 2381 | 9.188 | 1.867 | 3.605 | 17.33 | 9.34 |
| 10.0 | 213695 | 2519 | 2388 | 9.222 | 1.867 | 3.540 | 17.60 | 9.46 |
| 15.0 | 217477 | 2528 | 2395 | 9.254 | 1.868 | 3.478 | 17.88 | 9.59 |
| 20.0 | 221258 | 2537 | 2402 | 9.286 | 1.868 | 3.417 | 18.17 | 9.73 |
| 25.0 | 225039 | 2547 | 2409 | 9.318 | 1.869 | 3.359 | 18.47 | 9.87 |
| 30.0 | 228819 | 2556 | 2416 | 9.349 | 1.869 | 3.304 | 18.78 | 10.02 |
| 35.0 | 232598 | 2565 | 2423 | 9.380 | 1.870 | 3.249 | 19.10 | 10.17 |
| 40.0 | 236377 | 2575 | 2430 | 9.410 | 1.871 | 3.197 | 19.43 | 10.32 |
| 45.0 | 240155 | 2584 | 2437 | 9.439 | 1.872 | 3.147 | 19.77 | 10.47 |
| 50.0 | 243933 | 2593 | 2444 | 9.469 | 1.874 | 3.098 | 20.11 | 10.63 |
| 60.0 | 251489 | 2612 | 2459 | 9.526 | 1.876 | 3.004 | 20.82 | 10.96 |
| 70.0 | 259043 | 2631 | 2473 | 9.581 | 1.880 | 2.916 | 21.56 | 11.29 |
| 80.0 | 266597 | 2650 | 2487 | 9.635 | 1.883 | 2.833 | 22.31 | 11.64 |
| 90.0 | 274150 | 2669 | 2501 | 9.688 | 1.887 | 2.755 | 23.10 | 11.99 |
| 100.0 | 281703 | 2688 | 2515 | 9.739 | 1.891 | 2.681 | 23.90 | 12.35 |
| 125.0 | 300583 | 2735 | 2551 | 9.862 | 1.902 | 2.512 | 25.99 | 13.28 |
| 150.0 | 319462 | 2783 | 2587 | 9.978 | 1.914 | 2.364 | 28.19 | 14.23 |
| 175.0 | 338339 | 2831 | 2624 | 10.089 | 1.927 | 2.232 | 30.50 | 15.21 |
| 200.0 | 357216 | 2879 | 2661 | 10.194 | 1.940 | 2.114 | 32.89 | 16.21 |
| 225.0 | 376093 | 2928 | 2698 | 10.294 | 1.954 | 2.008 | 35.37 | 17.22 |
| 250.0 | 394969 | 2977 | 2735 | 10.390 | 1.969 | 1.912 | 37.93 | 18.24 |
| 275.0 | 413845 | 3026 | 2773 | 10.482 | 1.984 | 1.824 | 40.56 | 19.27 |
| 300.0 | 432721 | 3076 | 2811 | 10.571 | 2.000 | 1.745 | 43.26 | 20.30 |
| 325.0 | 451597 | 3126 | 2850 | 10.657 | 2.015 | 1.672 | 46.02 | 21.34 |
| 350.0 | 470473 | 3177 | 2889 | 10.739 | 2.031 | 1.605 | 48.84 | 22.38 |
| 375.0 | 489348 | 3228 | 2929 | 10.820 | 2.047 | 1.543 | 51.73 | 23.41 |
| 400.0 | 508224 | 3279 | 2968 | 10.897 | 2.064 | 1.486 | 54.66 | 24.45 |
| 450.0 | 545974 | 3383 | 3049 | 11.046 | 2.097 | 1.383 | 60.69 | 26.52 |
| 500.0 | 583725 | 3489 | 3132 | 11.188 | 2.131 | 1.293 | 66.90 | 28.57 |
| 550.0 | 621476 | 3596 | 3216 | 11.322 | 2.166 | 1.215 | 73.29 | 30.60 |
| 600.0 | 659226 | 3705 | 3302 | 11.451 | 2.201 | 1.145 | 79.83 | 32.61 |
| 650.0 | 696976 | 3816 | 3390 | 11.575 | 2.236 | 1.083 | 86.51 | 34.59 |
| 700.0 | 734726 | 3929 | 3480 | 11.694 | 2.272 | 1.028 | 93.32 | 36.55 |
| 750.0 | 772477 | 4043 | 3571 | 11.808 | 2.307 | 0.977 | 100.2 | 38.47 |
| 800.0 | 810227 | 4160 | 3664 | 11.919 | 2.342 | 0.932 | 107.3 | 40.37 |
| 850.0 | 847977 | 4278 | 3759 | 12.027 | 2.377 | 0.890 | 114.4 | 42.24 |
| 900.0 | 885727 | 4397 | 3856 | 12.131 | 2.411 | 0.852 | 121.6 | 44.07 |
| 950.0 | 923477 | 4519 | 3954 | 12.232 | 2.445 | 0.818 | 128.9 | 45.88 |
| 1000.0 | 961227 | 4642 | 4054 | 12.331 | 2.478 | 0.785 | 136.3 | 47.66 |

Table 1 (cont.)

$$p = 0.0020 \, \text{MPa} = 0.020 \ \text{bar}$$

| $t$ | $v$ | $h$ | $u$ | $s$ | $c_p$ | $\alpha_p$ | $\lambda$ | $\eta$ |
|---|---|---|---|---|---|---|---|---|
| °C | dm³/kg | kJ/kg | kJ/kg | kJ/kg K | kJ/kg K | $10^{-3}$/K | mW/K m | $\mu$Pa s |
| 0.0 | 1.0002 | −0.04 | −0.04 | −0.0002 | 4.229 | −0.081 | 561.0 | 1792 |
| 5.0 | 1.0001 | 21.0 | 21.0 | 0.076 | 4.200 | 0.011 | 570.5 | 1518 |
| 10.0 | 1.0003 | 42.0 | 42.0 | 0.151 | 4.188 | 0.087 | 580.0 | 1306 |
| 15.0 | 1.0009 | 62.9 | 62.9 | 0.224 | 4.184 | 0.152 | 589.3 | 1138 |
| 17.50 | saturation | | | | | | | |
| liquid | 1.0013 | 73.4 | 73.4 | 0.260 | 4.184 | 0.181 | 593.9 | 1066 |
| vapour | 66997 | 2532 | 2398 | 8.722 | 1.880 | 3.463 | 18.07 | 9.66 |
| 20.0 | 67578 | 2537 | 2402 | 8.738 | 1.879 | 3.432 | 18.22 | 9.73 |
| 25.0 | 68737 | 2546 | 2409 | 8.769 | 1.878 | 3.372 | 18.51 | 9.87 |
| 30.0 | 69895 | 2556 | 2416 | 8.801 | 1.878 | 3.315 | 18.82 | 10.01 |
| 35.0 | 71053 | 2565 | 2423 | 8.831 | 1.878 | 3.259 | 19.14 | 10.16 |
| 40.0 | 72211 | 2574 | 2430 | 8.862 | 1.878 | 3.206 | 19.47 | 10.32 |
| 45.0 | 73368 | 2584 | 2437 | 8.891 | 1.878 | 3.154 | 19.80 | 10.47 |
| 50.0 | 74525 | 2593 | 2444 | 8.921 | 1.879 | 3.104 | 20.14 | 10.63 |
| 60.0 | 76838 | 2612 | 2458 | 8.978 | 1.881 | 3.010 | 20.85 | 10.96 |
| 70.0 | 79150 | 2631 | 2472 | 9.034 | 1.883 | 2.921 | 21.58 | 11.29 |
| 80.0 | 81462 | 2650 | 2487 | 9.088 | 1.886 | 2.837 | 22.34 | 11.64 |
| 90.0 | 83773 | 2668 | 2501 | 9.140 | 1.889 | 2.758 | 23.11 | 11.99 |
| 100.0 | 86083 | 2687 | 2515 | 9.192 | 1.893 | 2.684 | 23.92 | 12.35 |
| 125.0 | 91857 | 2735 | 2551 | 9.315 | 1.903 | 2.514 | 26.00 | 13.28 |
| 150.0 | 97630 | 2783 | 2587 | 9.431 | 1.915 | 2.365 | 28.20 | 14.23 |
| 175.0 | 103402 | 2831 | 2624 | 9.541 | 1.927 | 2.233 | 30.50 | 15.21 |
| 200.0 | 109173 | 2879 | 2661 | 9.646 | 1.941 | 2.114 | 32.90 | 16.21 |
| 225.0 | 114943 | 2928 | 2698 | 9.747 | 1.955 | 2.008 | 35.38 | 17.22 |
| 250.0 | 120714 | 2977 | 2735 | 9.843 | 1.969 | 1.912 | 37.93 | 18.24 |
| 275.0 | 126484 | 3026 | 2773 | 9.935 | 1.984 | 1.825 | 40.56 | 19.27 |
| 300.0 | 132254 | 3076 | 2811 | 10.024 | 2.000 | 1.745 | 43.26 | 20.30 |
| 325.0 | 138024 | 3126 | 2850 | 10.110 | 2.015 | 1.672 | 46.02 | 21.34 |
| 350.0 | 143793 | 3177 | 2889 | 10.192 | 2.031 | 1.605 | 48.85 | 22.38 |
| 375.0 | 149563 | 3228 | 2929 | 10.273 | 2.047 | 1.543 | 51.73 | 23.41 |
| 400.0 | 155332 | 3279 | 2968 | 10.350 | 2.064 | 1.486 | 54.66 | 24.45 |
| 450.0 | 166871 | 3383 | 3049 | 10.499 | 2.097 | 1.383 | 60.69 | 26.52 |
| 500.0 | 178410 | 3489 | 3132 | 10.641 | 2.131 | 1.293 | 66.91 | 28.57 |
| 550.0 | 189948 | 3596 | 3216 | 10.775 | 2.166 | 1.215 | 73.29 | 30.60 |
| 600.0 | 201487 | 3705 | 3302 | 10.904 | 2.201 | 1.145 | 79.83 | 32.61 |
| 650.0 | 213025 | 3816 | 3390 | 11.028 | 2.236 | 1.083 | 86.51 | 34.59 |
| 700.0 | 224563 | 3929 | 3480 | 11.146 | 2.272 | 1.028 | 93.32 | 36.55 |
| 750.0 | 236102 | 4043 | 3571 | 11.261 | 2.307 | 0.977 | 100.2 | 38.47 |
| 800.0 | 247640 | 4160 | 3664 | 11.372 | 2.342 | 0.932 | 107.3 | 40.37 |
| 850.0 | 259178 | 4278 | 3759 | 11.480 | 2.377 | 0.890 | 114.4 | 42.24 |
| 900.0 | 270716 | 4397 | 3856 | 11.584 | 2.411 | 0.852 | 121.6 | 44.07 |
| 950.0 | 282254 | 4519 | 3954 | 11.685 | 2.445 | 0.818 | 128.9 | 45.88 |
| 1000.0 | 293792 | 4642 | 4054 | 11.784 | 2.478 | 0.785 | 136.3 | 47.66 |

Table 1 (cont.) $p = 0.0030\,\mathrm{MPa} = 0.030\ \mathrm{bar}$

| $t$ | $v$ | $h$ | $u$ | $s$ | $c_\mathrm{p}$ | $\alpha_\mathrm{p}$ | $\lambda$ | $\eta$ |
|---|---|---|---|---|---|---|---|---|
| °C | dm³/kg | kJ/kg | kJ/kg | kJ/kg K | kJ/kg K | $10^{-3}$/K | mW/K m | $\mu$Pa s |
| 0.0 | 1.0002 | −0.04 | −0.04 | −0.0002 | 4.229 | −0.081 | 561.0 | 1792 |
| 5.0 | 1.0001 | 21.0 | 21.0 | 0.076 | 4.200 | 0.011 | 570.5 | 1518 |
| 10.0 | 1.0003 | 42.0 | 42.0 | 0.151 | 4.188 | 0.087 | 580.0 | 1306 |
| 15.0 | 1.0009 | 62.9 | 62.9 | 0.224 | 4.184 | 0.152 | 589.3 | 1138 |
| 20.0 | 1.0018 | 83.8 | 83.8 | 0.296 | 4.183 | 0.209 | 598.4 | 1002 |
| 24.08 | | | | saturation | | | | |
| liquid | 1.0028 | 100.9 | 100.9 | 0.354 | 4.183 | 0.250 | 605.6 | 909.4 |
| vapour | 45661 | 2544 | 2407 | 8.576 | 1.886 | 3.392 | 18.49 | 9.84 |
| 25.0 | 45803 | 2546 | 2408 | 8.581 | 1.886 | 3.381 | 18.54 | 9.87 |
| 30.0 | 46577 | 2555 | 2416 | 8.613 | 1.884 | 3.322 | 18.85 | 10.01 |
| 35.0 | 47350 | 2565 | 2423 | 8.643 | 1.883 | 3.266 | 19.17 | 10.16 |
| 40.0 | 48123 | 2574 | 2430 | 8.674 | 1.883 | 3.212 | 19.49 | 10.31 |
| 45.0 | 48896 | 2584 | 2437 | 8.704 | 1.883 | 3.160 | 19.82 | 10.47 |
| 50.0 | 49669 | 2593 | 2444 | 8.733 | 1.883 | 3.109 | 20.16 | 10.63 |
| 60.0 | 51212 | 2612 | 2458 | 8.790 | 1.884 | 3.014 | 20.87 | 10.96 |
| 70.0 | 52755 | 2631 | 2472 | 8.846 | 1.886 | 2.924 | 21.60 | 11.29 |
| 80.0 | 54297 | 2649 | 2487 | 8.900 | 1.888 | 2.840 | 22.35 | 11.64 |
| 90.0 | 55839 | 2668 | 2501 | 8.953 | 1.891 | 2.760 | 23.13 | 11.99 |
| 100.0 | 57380 | 2687 | 2515 | 9.004 | 1.894 | 2.685 | 23.93 | 12.35 |
| 125.0 | 61231 | 2735 | 2551 | 9.128 | 1.904 | 2.515 | 26.01 | 13.27 |
| 150.0 | 65081 | 2782 | 2587 | 9.244 | 1.915 | 2.366 | 28.21 | 14.23 |
| 175.0 | 68930 | 2831 | 2624 | 9.354 | 1.928 | 2.233 | 30.51 | 15.21 |
| 200.0 | 72778 | 2879 | 2661 | 9.459 | 1.941 | 2.115 | 32.90 | 16.21 |
| 225.0 | 76626 | 2928 | 2698 | 9.559 | 1.955 | 2.008 | 35.38 | 17.22 |
| 250.0 | 80473 | 2977 | 2735 | 9.656 | 1.970 | 1.912 | 37.93 | 18.24 |
| 275.0 | 84320 | 3026 | 2773 | 9.748 | 1.985 | 1.825 | 40.56 | 19.27 |
| 300.0 | 88167 | 3076 | 2811 | 9.837 | 2.000 | 1.745 | 43.26 | 20.30 |
| 325.0 | 92014 | 3126 | 2850 | 9.922 | 2.016 | 1.672 | 46.02 | 21.34 |
| 350.0 | 95860 | 3177 | 2889 | 10.005 | 2.031 | 1.605 | 48.85 | 22.38 |
| 375.0 | 99707 | 3228 | 2929 | 10.085 | 2.047 | 1.543 | 51.73 | 23.41 |
| 400.0 | 103553 | 3279 | 2968 | 10.163 | 2.064 | 1.486 | 54.66 | 24.45 |
| 450.0 | 111246 | 3383 | 3049 | 10.312 | 2.097 | 1.383 | 60.69 | 26.52 |
| 500.0 | 118939 | 3489 | 3132 | 10.454 | 2.131 | 1.294 | 66.91 | 28.57 |
| 550.0 | 126631 | 3596 | 3216 | 10.588 | 2.166 | 1.215 | 73.29 | 30.60 |
| 600.0 | 134324 | 3705 | 3302 | 10.717 | 2.201 | 1.145 | 79.83 | 32.61 |
| 650.0 | 142016 | 3816 | 3390 | 10.840 | 2.236 | 1.083 | 86.51 | 34.59 |
| 700.0 | 149708 | 3929 | 3480 | 10.959 | 2.272 | 1.028 | 93.32 | 36.55 |
| 750.0 | 157400 | 4043 | 3571 | 11.074 | 2.307 | 0.977 | 100.2 | 38.47 |
| 800.0 | 165093 | 4160 | 3664 | 11.185 | 2.342 | 0.932 | 107.3 | 40.37 |
| 850.0 | 172785 | 4278 | 3759 | 11.292 | 2.377 | 0.890 | 114.4 | 42.24 |
| 900.0 | 180477 | 4397 | 3856 | 11.397 | 2.411 | 0.852 | 121.6 | 44.07 |
| 950.0 | 188169 | 4519 | 3954 | 11.498 | 2.445 | 0.818 | 128.9 | 45.88 |
| 1000.0 | 195861 | 4642 | 4054 | 11.597 | 2.478 | 0.785 | 136.3 | 47.66 |

| $t$ | $v$ | $h$ | $u$ | $s$ | $c_\mathrm{p}$ | $\alpha_\mathrm{p}$ | $\lambda$ | $\eta$ |
|---|---|---|---|---|---|---|---|---|
| °C | dm³/kg | kJ/kg | kJ/kg | kJ/kg K | kJ/kg K | $10^{-3}$/K | mW/K m | $\mu$Pa s |
| 0.0 | 1.0002 | −0.04 | −0.04 | −0.0002 | 4.229 | −0.081 | 561.0 | 1792 |
| 5.0 | 1.0001 | 21.0 | 21.0 | 0.076 | 4.200 | 0.011 | 570.5 | 1518 |
| 10.0 | 1.0003 | 42.0 | 42.0 | 0.151 | 4.188 | 0.087 | 580.0 | 1306 |
| 15.0 | 1.0009 | 62.9 | 62.9 | 0.224 | 4.184 | 0.152 | 589.3 | 1138 |
| 20.0 | 1.0018 | 83.8 | 83.8 | 0.296 | 4.183 | 0.209 | 598.4 | 1002 |
| 25.0 | 1.0030 | 104.8 | 104.7 | 0.367 | 4.183 | 0.259 | 607.1 | 890.5 |
| 28.97 | | | | saturation | | | | |
| liquid | 1.0041 | 121.3 | 121.3 | 0.422 | 4.183 | 0.296 | 613.7 | 815.5 |
| vapour | 34797 | 2553 | 2414 | 8.473 | 1.891 | 3.343 | 18.81 | 9.98 |
| 30.0 | 34918 | 2555 | 2415 | 8.479 | 1.891 | 3.330 | 18.88 | 10.01 |
| 35.0 | 35499 | 2564 | 2422 | 8.510 | 1.889 | 3.273 | 19.19 | 10.16 |
| 40.0 | 36080 | 2574 | 2429 | 8.540 | 1.888 | 3.218 | 19.51 | 10.31 |
| 45.0 | 36660 | 2583 | 2437 | 8.570 | 1.887 | 3.165 | 19.85 | 10.47 |
| 50.0 | 37240 | 2593 | 2444 | 8.600 | 1.887 | 3.114 | 20.18 | 10.63 |
| 60.0 | 38399 | 2612 | 2458 | 8.657 | 1.887 | 3.017 | 20.88 | 10.95 |
| 70.0 | 39558 | 2630 | 2472 | 8.713 | 1.888 | 2.927 | 21.61 | 11.29 |
| 80.0 | 40715 | 2649 | 2486 | 8.767 | 1.890 | 2.842 | 22.36 | 11.64 |
| 90.0 | 41872 | 2668 | 2501 | 8.820 | 1.892 | 2.762 | 23.14 | 11.99 |
| 100.0 | 43029 | 2687 | 2515 | 8.871 | 1.895 | 2.687 | 23.94 | 12.35 |
| 125.0 | 45918 | 2735 | 2551 | 8.995 | 1.905 | 2.517 | 26.02 | 13.27 |
| 150.0 | 48807 | 2782 | 2587 | 9.111 | 1.916 | 2.367 | 28.22 | 14.23 |
| 175.0 | 51694 | 2830 | 2624 | 9.221 | 1.928 | 2.234 | 30.51 | 15.21 |
| 200.0 | 54580 | 2879 | 2661 | 9.326 | 1.941 | 2.115 | 32.91 | 16.21 |
| 225.0 | 57467 | 2928 | 2698 | 9.427 | 1.955 | 2.009 | 35.38 | 17.22 |
| 250.0 | 60352 | 2977 | 2735 | 9.523 | 1.970 | 1.913 | 37.94 | 18.24 |
| 275.0 | 63238 | 3026 | 2773 | 9.615 | 1.985 | 1.825 | 40.56 | 19.27 |
| 300.0 | 66123 | 3076 | 2811 | 9.704 | 2.000 | 1.745 | 43.26 | 20.30 |
| 325.0 | 69009 | 3126 | 2850 | 9.790 | 2.016 | 1.672 | 46.02 | 21.34 |
| 350.0 | 71894 | 3177 | 2889 | 9.872 | 2.031 | 1.605 | 48.85 | 22.38 |
| 375.0 | 74779 | 3228 | 2928 | 9.953 | 2.048 | 1.543 | 51.73 | 23.41 |
| 400.0 | 77664 | 3279 | 2968 | 10.030 | 2.064 | 1.486 | 54.67 | 24.45 |
| 450.0 | 83434 | 3383 | 3049 | 10.179 | 2.097 | 1.383 | 60.69 | 26.52 |
| 500.0 | 89203 | 3489 | 3132 | 10.321 | 2.131 | 1.294 | 66.91 | 28.57 |
| 550.0 | 94973 | 3596 | 3216 | 10.455 | 2.166 | 1.215 | 73.29 | 30.60 |
| 600.0 | 100742 | 3705 | 3302 | 10.584 | 2.201 | 1.145 | 79.83 | 32.61 |
| 650.0 | 106511 | 3816 | 3390 | 10.708 | 2.236 | 1.083 | 86.51 | 34.59 |
| 700.0 | 112281 | 3929 | 3480 | 10.827 | 2.272 | 1.028 | 93.32 | 36.55 |
| 750.0 | 118050 | 4043 | 3571 | 10.941 | 2.307 | 0.977 | 100.2 | 38.47 |
| 800.0 | 123819 | 4160 | 3664 | 11.052 | 2.342 | 0.932 | 107.3 | 40.37 |
| 850.0 | 129588 | 4278 | 3759 | 11.160 | 2.377 | 0.890 | 114.4 | 42.24 |
| 900.0 | 135358 | 4397 | 3856 | 11.264 | 2.411 | 0.852 | 121.6 | 44.07 |
| 950.0 | 141127 | 4519 | 3954 | 11.365 | 2.445 | 0.818 | 128.9 | 45.88 |
| 1000.0 | 146896 | 4642 | 4054 | 11.464 | 2.478 | 0.785 | 136.3 | 47.66 |

Table 1 (cont.)

$$p = 0.0050 \, \text{MPa} = 0.050 \, \text{bar}$$

| $t$ | $v$ | $h$ | $u$ | $s$ | $c_\text{p}$ | $\alpha_\text{p}$ | $\lambda$ | $\eta$ |
|---|---|---|---|---|---|---|---|---|
| °C | dm³/kg | kJ/kg | kJ/kg | kJ/kg K | kJ/kg K | $10^{-3}$/K | mW/K m | $\mu$Pa s |
| 0.0 | 1.0002 | −0.04 | −0.04 | −0.0002 | 4.229 | −0.081 | 561.0 | 1792 |
| 5.0 | 1.0001 | 21.0 | 21.0 | 0.076 | 4.200 | 0.011 | 570.5 | 1518 |
| 10.0 | 1.0003 | 42.0 | 42.0 | 0.151 | 4.188 | 0.087 | 580.0 | 1306 |
| 15.0 | 1.0009 | 62.9 | 62.9 | 0.224 | 4.184 | 0.152 | 589.3 | 1138 |
| 20.0 | 1.0018 | 83.8 | 83.8 | 0.296 | 4.183 | 0.209 | 598.4 | 1002 |
| 25.0 | 1.0030 | 104.8 | 104.7 | 0.367 | 4.183 | 0.259 | 607.1 | 890.5 |
| 30.0 | 1.0044 | 125.7 | 125.7 | 0.437 | 4.183 | 0.305 | 615.4 | 797.7 |
| 32.88 | | | | saturation | | | | |
| liquid | 1.0053 | 137.7 | 137.7 | 0.476 | 4.183 | 0.330 | 620.0 | 751.1 |
| vapour | 28191 | 2560 | 2419 | 8.393 | 1.895 | 3.305 | 19.08 | 10.09 |
| 35.0 | 28388 | 2564 | 2422 | 8.406 | 1.894 | 3.280 | 19.22 | 10.16 |
| 40.0 | 28853 | 2574 | 2429 | 8.437 | 1.893 | 3.224 | 19.54 | 10.31 |
| 45.0 | 29318 | 2583 | 2436 | 8.467 | 1.891 | 3.171 | 19.87 | 10.47 |
| 50.0 | 29783 | 2592 | 2444 | 8.496 | 1.891 | 3.119 | 20.21 | 10.63 |
| 60.0 | 30711 | 2611 | 2458 | 8.554 | 1.890 | 3.021 | 20.90 | 10.95 |
| 70.0 | 31639 | 2630 | 2472 | 8.610 | 1.891 | 2.930 | 21.63 | 11.29 |
| 80.0 | 32566 | 2649 | 2486 | 8.664 | 1.892 | 2.845 | 22.38 | 11.63 |
| 90.0 | 33492 | 2668 | 2501 | 8.717 | 1.894 | 2.765 | 23.15 | 11.99 |
| 100.0 | 34418 | 2687 | 2515 | 8.768 | 1.897 | 2.689 | 23.95 | 12.35 |
| 125.0 | 36731 | 2735 | 2551 | 8.891 | 1.906 | 2.518 | 26.03 | 13.27 |
| 150.0 | 39042 | 2782 | 2587 | 9.008 | 1.916 | 2.367 | 28.22 | 14.23 |
| 175.0 | 41352 | 2830 | 2624 | 9.118 | 1.929 | 2.234 | 30.52 | 15.21 |
| 200.0 | 43662 | 2879 | 2660 | 9.223 | 1.942 | 2.116 | 32.91 | 16.21 |
| 225.0 | 45971 | 2927 | 2698 | 9.324 | 1.956 | 2.009 | 35.38 | 17.22 |
| 250.0 | 48280 | 2977 | 2735 | 9.420 | 1.970 | 1.913 | 37.94 | 18.24 |
| 275.0 | 50589 | 3026 | 2773 | 9.512 | 1.985 | 1.825 | 40.57 | 19.27 |
| 300.0 | 52897 | 3076 | 2811 | 9.601 | 2.000 | 1.746 | 43.26 | 20.30 |
| 325.0 | 55206 | 3126 | 2850 | 9.687 | 2.016 | 1.672 | 46.03 | 21.34 |
| 350.0 | 57514 | 3177 | 2889 | 9.769 | 2.032 | 1.605 | 48.85 | 22.38 |
| 375.0 | 59822 | 3228 | 2928 | 9.850 | 2.048 | 1.543 | 51.73 | 23.41 |
| 400.0 | 62130 | 3279 | 2968 | 9.927 | 2.064 | 1.486 | 54.67 | 24.45 |
| 450.0 | 66746 | 3383 | 3049 | 10.076 | 2.097 | 1.383 | 60.69 | 26.52 |
| 500.0 | 71362 | 3489 | 3132 | 10.218 | 2.131 | 1.294 | 66.91 | 28.57 |
| 550.0 | 75978 | 3596 | 3216 | 10.352 | 2.166 | 1.215 | 73.29 | 30.60 |
| 600.0 | 80593 | 3705 | 3302 | 10.481 | 2.201 | 1.145 | 79.83 | 32.61 |
| 650.0 | 85209 | 3816 | 3390 | 10.605 | 2.236 | 1.083 | 86.51 | 34.59 |
| 700.0 | 89824 | 3929 | 3480 | 10.724 | 2.272 | 1.028 | 93.32 | 36.55 |
| 750.0 | 94440 | 4043 | 3571 | 10.838 | 2.307 | 0.977 | 100.2 | 38.47 |
| 800.0 | 99055 | 4160 | 3664 | 10.949 | 2.342 | 0.932 | 107.3 | 40.37 |
| 850.0 | 103670 | 4278 | 3759 | 11.057 | 2.377 | 0.890 | 114.4 | 42.24 |
| 900.0 | 108286 | 4397 | 3856 | 11.161 | 2.411 | 0.852 | 121.6 | 44.07 |
| 950.0 | 112901 | 4519 | 3954 | 11.262 | 2.445 | 0.818 | 128.9 | 45.88 |
| 1000.0 | 117516 | 4642 | 4054 | 11.361 | 2.478 | 0.785 | 136.3 | 47.66 |

Table 1 (cont.)

$p = 0.0075\,\text{MPa} = 0.075$ bar

| $t$ | $v$ | $h$ | $u$ | $s$ | $c_\text{p}$ | $\alpha_\text{p}$ | $\lambda$ | $\eta$ |
|------|--------|--------|--------|---------|---------|---------|---------|---------|
| °C | dm$^3$/kg | kJ/kg | kJ/kg | kJ/kg K | kJ/kg K | 10$^{-3}$/K | mW/K m | $\mu$Pa s |
| 0.0 | 1.0002 | −0.03 | −0.04 | −0.0002 | 4.229 | −0.081 | 561.0 | 1792 |
| 5.0 | 1.0001 | 21.0 | 21.0 | 0.076 | 4.200 | 0.011 | 570.5 | 1518 |
| 10.0 | 1.0003 | 42.0 | 42.0 | 0.151 | 4.188 | 0.087 | 580.0 | 1306 |
| 15.0 | 1.0009 | 62.9 | 62.9 | 0.224 | 4.184 | 0.152 | 589.3 | 1138 |
| 20.0 | 1.0018 | 83.8 | 83.8 | 0.296 | 4.183 | 0.209 | 598.4 | 1002 |
| 25.0 | 1.0030 | 104.8 | 104.7 | 0.367 | 4.183 | 0.259 | 607.1 | 890.5 |
| 30.0 | 1.0044 | 125.7 | 125.7 | 0.437 | 4.183 | 0.305 | 615.4 | 797.7 |
| 35.0 | 1.0060 | 146.6 | 146.6 | 0.505 | 4.183 | 0.347 | 623.2 | 719.6 |
| 40.0 | 1.0079 | 167.5 | 167.5 | 0.572 | 4.182 | 0.386 | 630.5 | 653.2 |
| 40.30 | | | | saturation | | | | |
| liquid | 1.0080 | 168.8 | 168.7 | 0.576 | 4.182 | 0.388 | 631.0 | 649.6 |
| vapour | 19237 | 2573 | 2429 | 8.249 | 1.905 | 3.237 | 19.62 | 10.32 |
| 45.0 | 19529 | 2582 | 2436 | 8.278 | 1.902 | 3.185 | 19.93 | 10.46 |
| 50.0 | 19840 | 2592 | 2443 | 8.307 | 1.900 | 3.132 | 20.26 | 10.62 |
| 60.0 | 20461 | 2611 | 2457 | 8.365 | 1.898 | 3.032 | 20.95 | 10.95 |
| 70.0 | 21081 | 2630 | 2472 | 8.421 | 1.897 | 2.938 | 21.67 | 11.29 |
| 80.0 | 21700 | 2649 | 2486 | 8.476 | 1.897 | 2.852 | 22.42 | 11.63 |
| 90.0 | 22318 | 2668 | 2500 | 8.529 | 1.898 | 2.770 | 23.19 | 11.99 |
| 100.0 | 22937 | 2687 | 2515 | 8.580 | 1.900 | 2.694 | 23.98 | 12.35 |
| 125.0 | 24480 | 2734 | 2551 | 8.704 | 1.908 | 2.521 | 26.05 | 13.27 |
| 150.0 | 26022 | 2782 | 2587 | 8.820 | 1.918 | 2.370 | 28.24 | 14.23 |
| 175.0 | 27564 | 2830 | 2624 | 8.931 | 1.930 | 2.236 | 30.53 | 15.21 |
| 200.0 | 29104 | 2879 | 2660 | 9.036 | 1.943 | 2.117 | 32.92 | 16.20 |
| 225.0 | 30644 | 2927 | 2698 | 9.136 | 1.956 | 2.010 | 35.39 | 17.22 |
| 250.0 | 32184 | 2976 | 2735 | 9.232 | 1.971 | 1.913 | 37.94 | 18.24 |
| 275.0 | 33723 | 3026 | 2773 | 9.325 | 1.985 | 1.826 | 40.57 | 19.27 |
| 300.0 | 35263 | 3076 | 2811 | 9.414 | 2.000 | 1.746 | 43.27 | 20.30 |
| 325.0 | 36802 | 3126 | 2850 | 9.499 | 2.016 | 1.673 | 46.03 | 21.34 |
| 350.0 | 38341 | 3177 | 2889 | 9.582 | 2.032 | 1.606 | 48.85 | 22.38 |
| 375.0 | 39880 | 3228 | 2928 | 9.662 | 2.048 | 1.544 | 51.73 | 23.41 |
| 400.0 | 41419 | 3279 | 2968 | 9.740 | 2.064 | 1.486 | 54.67 | 24.45 |
| 450.0 | 44496 | 3383 | 3049 | 9.889 | 2.097 | 1.383 | 60.69 | 26.52 |
| 500.0 | 47574 | 3489 | 3132 | 10.031 | 2.131 | 1.294 | 66.91 | 28.57 |
| 550.0 | 50651 | 3596 | 3216 | 10.165 | 2.166 | 1.215 | 73.29 | 30.60 |
| 600.0 | 53728 | 3705 | 3302 | 10.294 | 2.201 | 1.145 | 79.83 | 32.61 |
| 650.0 | 56805 | 3816 | 3390 | 10.418 | 2.236 | 1.083 | 86.51 | 34.59 |
| 700.0 | 59882 | 3929 | 3480 | 10.536 | 2.272 | 1.028 | 93.32 | 36.55 |
| 750.0 | 62959 | 4043 | 3571 | 10.651 | 2.307 | 0.977 | 100.2 | 38.47 |
| 800.0 | 66036 | 4160 | 3664 | 10.762 | 2.342 | 0.932 | 107.3 | 40.37 |
| 850.0 | 69113 | 4278 | 3759 | 10.869 | 2.377 | 0.890 | 114.4 | 42.24 |
| 900.0 | 72190 | 4397 | 3856 | 10.974 | 2.411 | 0.852 | 121.6 | 44.07 |
| 950.0 | 75267 | 4519 | 3954 | 11.075 | 2.445 | 0.818 | 128.9 | 45.88 |
| 1000.0 | 78344 | 4642 | 4054 | 11.174 | 2.478 | 0.785 | 136.3 | 47.66 |

Table 1 (cont.)

$$p = 0.01\,\text{MPa} = 0.1\ \text{bar}$$

| $t$ | $v$ | $h$ | $u$ | $s$ | $c_p$ | $\alpha_p$ | $\lambda$ | $\eta$ |
|------|------|------|------|------|------|------|------|------|
| °C | dm³/kg | kJ/kg | kJ/kg | kJ/kg K | kJ/kg K | $10^{-3}$/K | mW/K m | $\mu$Pa s |
| 0.0 | 1.0002 | −0.03 | −0.04 | −0.0002 | 4.229 | −0.081 | 561.0 | 1792 |
| 5.0 | 1.0001 | 21.0 | 21.0 | 0.076 | 4.200 | 0.011 | 570.5 | 1518 |
| 10.0 | 1.0003 | 42.0 | 42.0 | 0.151 | 4.188 | 0.087 | 580.0 | 1306 |
| 15.0 | 1.0009 | 62.9 | 62.9 | 0.224 | 4.184 | 0.152 | 589.3 | 1138 |
| 20.0 | 1.0018 | 83.8 | 83.8 | 0.296 | 4.183 | 0.209 | 598.4 | 1002 |
| 25.0 | 1.0030 | 104.8 | 104.7 | 0.367 | 4.183 | 0.259 | 607.1 | 890.5 |
| 30.0 | 1.0044 | 125.7 | 125.7 | 0.437 | 4.183 | 0.305 | 615.4 | 797.7 |
| 35.0 | 1.0060 | 146.6 | 146.6 | 0.505 | 4.183 | 0.347 | 623.2 | 719.6 |
| 40.0 | 1.0079 | 167.5 | 167.5 | 0.572 | 4.182 | 0.386 | 630.5 | 653.2 |
| 45.0 | 1.0099 | 188.4 | 188.4 | 0.639 | 4.182 | 0.423 | 637.3 | 596.3 |
| 45.82 | | | | saturation | | | | |
| liquid | 1.0103 | 191.8 | 191.8 | 0.649 | 4.182 | 0.428 | 638.4 | 587.8 |
| vapour | 14673 | 2583 | 2437 | 8.148 | 1.913 | 3.190 | 20.04 | 10.49 |
| 50.0 | 14869 | 2591 | 2443 | 8.173 | 1.910 | 3.144 | 20.31 | 10.62 |
| 60.0 | 15336 | 2610 | 2457 | 8.231 | 1.906 | 3.042 | 21.00 | 10.95 |
| 70.0 | 15802 | 2629 | 2471 | 8.288 | 1.903 | 2.947 | 21.71 | 11.28 |
| 80.0 | 16267 | 2648 | 2486 | 8.342 | 1.902 | 2.858 | 22.45 | 11.63 |
| 90.0 | 16732 | 2667 | 2500 | 8.395 | 1.903 | 2.776 | 23.22 | 11.98 |
| 100.0 | 17196 | 2686 | 2514 | 8.447 | 1.904 | 2.698 | 24.01 | 12.34 |
| 125.0 | 18355 | 2734 | 2551 | 8.571 | 1.910 | 2.524 | 26.07 | 13.27 |
| 150.0 | 19513 | 2782 | 2587 | 8.687 | 1.920 | 2.372 | 28.26 | 14.23 |
| 175.0 | 20669 | 2830 | 2623 | 8.798 | 1.931 | 2.238 | 30.54 | 15.21 |
| 200.0 | 21825 | 2879 | 2660 | 8.903 | 1.943 | 2.118 | 32.93 | 16.20 |
| 225.0 | 22981 | 2927 | 2697 | 9.003 | 1.957 | 2.011 | 35.40 | 17.21 |
| 250.0 | 24136 | 2976 | 2735 | 9.100 | 1.971 | 1.914 | 37.95 | 18.24 |
| 275.0 | 25291 | 3026 | 2773 | 9.192 | 1.986 | 1.826 | 40.58 | 19.27 |
| 300.0 | 26445 | 3076 | 2811 | 9.281 | 2.001 | 1.746 | 43.27 | 20.30 |
| 325.0 | 27600 | 3126 | 2850 | 9.367 | 2.016 | 1.673 | 46.03 | 21.34 |
| 350.0 | 28754 | 3177 | 2889 | 9.449 | 2.032 | 1.606 | 48.85 | 22.38 |
| 375.0 | 29909 | 3228 | 2928 | 9.530 | 2.048 | 1.544 | 51.74 | 23.41 |
| 400.0 | 31063 | 3279 | 2968 | 9.607 | 2.064 | 1.486 | 54.67 | 24.45 |
| 450.0 | 33371 | 3383 | 3049 | 9.757 | 2.097 | 1.383 | 60.70 | 26.52 |
| 500.0 | 35679 | 3489 | 3132 | 9.898 | 2.131 | 1.294 | 66.91 | 28.57 |
| 550.0 | 37987 | 3596 | 3216 | 10.032 | 2.166 | 1.215 | 73.30 | 30.60 |
| 600.0 | 40295 | 3705 | 3302 | 10.161 | 2.201 | 1.145 | 79.83 | 32.61 |
| 650.0 | 42603 | 3816 | 3390 | 10.285 | 2.236 | 1.083 | 86.51 | 34.59 |
| 700.0 | 44911 | 3929 | 3480 | 10.404 | 2.272 | 1.028 | 93.32 | 36.55 |
| 750.0 | 47219 | 4043 | 3571 | 10.518 | 2.307 | 0.977 | 100.2 | 38.47 |
| 800.0 | 49527 | 4160 | 3664 | 10.629 | 2.342 | 0.932 | 107.3 | 40.37 |
| 850.0 | 51835 | 4278 | 3759 | 10.737 | 2.377 | 0.890 | 114.4 | 42.24 |
| 900.0 | 54142 | 4397 | 3856 | 10.841 | 2.411 | 0.852 | 121.6 | 44.07 |
| 950.0 | 56450 | 4519 | 3954 | 10.942 | 2.445 | 0.818 | 128.9 | 45.88 |
| 1000.0 | 58758 | 4642 | 4054 | 11.041 | 2.478 | 0.785 | 136.3 | 47.66 |

Table 1 (cont.)

$$p = 0.02\,\text{MPa} = 0.2 \ \text{bar}$$

| $t$ | $v$ | $h$ | $u$ | $s$ | $c_p$ | $\alpha_p$ | $\lambda$ | $\eta$ |
|---|---|---|---|---|---|---|---|---|
| °C | dm³/kg | kJ/kg | kJ/kg | kJ/kg K | kJ/kg K | $10^{-3}$/K | mW/K m | $\mu$Pa s |
| 0.0 | 1.0002 | −0.02 | −0.04 | −0.0002 | 4.229 | −0.081 | 561.0 | 1792 |
| 5.0 | 1.0000 | 21.0 | 21.0 | 0.076 | 4.200 | 0.011 | 570.5 | 1518 |
| 10.0 | 1.0003 | 42.0 | 42.0 | 0.151 | 4.188 | 0.087 | 580.0 | 1306 |
| 15.0 | 1.0009 | 62.9 | 62.9 | 0.224 | 4.184 | 0.152 | 589.3 | 1138 |
| 20.0 | 1.0018 | 83.9 | 83.8 | 0.296 | 4.183 | 0.209 | 598.4 | 1002 |
| 25.0 | 1.0030 | 104.8 | 104.7 | 0.367 | 4.183 | 0.259 | 607.1 | 890.5 |
| 30.0 | 1.0044 | 125.7 | 125.7 | 0.437 | 4.183 | 0.305 | 615.4 | 797.7 |
| 35.0 | 1.0060 | 146.6 | 146.6 | 0.505 | 4.183 | 0.347 | 623.2 | 719.6 |
| 40.0 | 1.0079 | 167.5 | 167.5 | 0.572 | 4.182 | 0.386 | 630.5 | 653.2 |
| 45.0 | 1.0099 | 188.4 | 188.4 | 0.639 | 4.182 | 0.423 | 637.3 | 596.3 |
| 50.0 | 1.0122 | 209.3 | 209.3 | 0.704 | 4.182 | 0.457 | 643.5 | 547.1 |
| 60.0 | 1.0171 | 251.2 | 251.1 | 0.831 | 4.183 | 0.522 | 654.3 | 466.6 |
| 60.07 | | | | saturation | | | | |
| liquid | 1.0172 | 251.5 | 251.4 | 0.832 | 4.183 | 0.523 | 654.4 | 466.1 |
| vapour | 7649.9 | 2608 | 2455 | 7.907 | 1.937 | 3.082 | 21.19 | 10.94 |
| 70.0 | 7883.5 | 2628 | 2470 | 7.964 | 1.929 | 2.980 | 21.88 | 11.27 |
| 80.0 | 8118.1 | 2647 | 2484 | 8.019 | 1.923 | 2.886 | 22.60 | 11.62 |
| 90.0 | 8352.0 | 2666 | 2499 | 8.073 | 1.920 | 2.798 | 23.35 | 11.97 |
| 100.0 | 8585.5 | 2685 | 2514 | 8.125 | 1.918 | 2.717 | 24.12 | 12.34 |
| 125.0 | 9167.7 | 2733 | 2550 | 8.249 | 1.920 | 2.536 | 26.16 | 13.26 |
| 150.0 | 9748.4 | 2781 | 2586 | 8.366 | 1.926 | 2.380 | 28.32 | 14.22 |
| 175.0 | 10328 | 2830 | 2623 | 8.477 | 1.936 | 2.244 | 30.59 | 15.20 |
| 200.0 | 10907 | 2878 | 2660 | 8.582 | 1.947 | 2.123 | 32.97 | 16.20 |
| 225.0 | 11485 | 2927 | 2697 | 8.683 | 1.960 | 2.014 | 35.43 | 17.21 |
| 250.0 | 12063 | 2976 | 2735 | 8.779 | 1.973 | 1.917 | 37.97 | 18.23 |
| 275.0 | 12641 | 3026 | 2773 | 8.872 | 1.987 | 1.828 | 40.60 | 19.26 |
| 300.0 | 13219 | 3075 | 2811 | 8.961 | 2.002 | 1.748 | 43.29 | 20.30 |
| 325.0 | 13797 | 3126 | 2850 | 9.046 | 2.017 | 1.674 | 46.05 | 21.34 |
| 350.0 | 14374 | 3176 | 2889 | 9.129 | 2.033 | 1.607 | 48.87 | 22.37 |
| 375.0 | 14952 | 3227 | 2928 | 9.210 | 2.049 | 1.545 | 51.75 | 23.41 |
| 400.0 | 15529 | 3279 | 2968 | 9.287 | 2.065 | 1.487 | 54.68 | 24.45 |
| 450.0 | 16684 | 3383 | 3049 | 9.436 | 2.098 | 1.384 | 60.70 | 26.52 |
| 500.0 | 17838 | 3489 | 3132 | 9.578 | 2.132 | 1.294 | 66.92 | 28.57 |
| 550.0 | 18992 | 3596 | 3216 | 9.712 | 2.166 | 1.215 | 73.30 | 30.60 |
| 600.0 | 20147 | 3705 | 3302 | 9.841 | 2.201 | 1.146 | 79.84 | 32.61 |
| 650.0 | 21301 | 3816 | 3390 | 9.965 | 2.236 | 1.084 | 86.52 | 34.59 |
| 700.0 | 22455 | 3929 | 3480 | 10.084 | 2.272 | 1.028 | 93.33 | 36.55 |
| 750.0 | 23609 | 4043 | 3571 | 10.198 | 2.307 | 0.978 | 100.3 | 38.47 |
| 800.0 | 24763 | 4160 | 3664 | 10.309 | 2.343 | 0.932 | 107.3 | 40.37 |
| 850.0 | 25917 | 4278 | 3759 | 10.417 | 2.377 | 0.890 | 114.4 | 42.24 |
| 900.0 | 27071 | 4397 | 3856 | 10.521 | 2.412 | 0.853 | 121.6 | 44.07 |
| 950.0 | 28225 | 4519 | 3954 | 10.622 | 2.445 | 0.818 | 128.9 | 45.88 |
| 1000.0 | 29378 | 4642 | 4054 | 10.721 | 2.478 | 0.786 | 136.3 | 47.66 |

Table 1 (cont.) $p = 0.03\,\text{MPa} = 0.3\ \text{bar}$

| $t$ | $v$ | $h$ | $u$ | $s$ | $c_p$ | $\alpha_p$ | $\lambda$ | $\eta$ |
|------|------|------|------|------|------|------|------|------|
| °C | dm³/kg | kJ/kg | kJ/kg | kJ/kg K | kJ/kg K | $10^{-3}$/K | mW/K m | $\mu$Pa s |
| 0.0 | 1.0002 | −0.01 | −0.04 | −0.0002 | 4.229 | −0.081 | 561.0 | 1792 |
| 5.0 | 1.0000 | 21.1 | 21.0 | 0.076 | 4.200 | 0.011 | 570.5 | 1518 |
| 10.0 | 1.0003 | 42.0 | 42.0 | 0.151 | 4.188 | 0.087 | 580.0 | 1306 |
| 15.0 | 1.0009 | 62.9 | 62.9 | 0.224 | 4.184 | 0.152 | 589.3 | 1138 |
| 20.0 | 1.0018 | 83.9 | 83.8 | 0.296 | 4.183 | 0.209 | 598.4 | 1002 |
| 25.0 | 1.0030 | 104.8 | 104.7 | 0.367 | 4.183 | 0.259 | 607.1 | 890.5 |
| 30.0 | 1.0044 | 125.7 | 125.7 | 0.437 | 4.183 | 0.305 | 615.4 | 797.7 |
| 35.0 | 1.0060 | 146.6 | 146.6 | 0.505 | 4.183 | 0.347 | 623.3 | 719.6 |
| 40.0 | 1.0079 | 167.5 | 167.5 | 0.572 | 4.182 | 0.386 | 630.6 | 653.2 |
| 45.0 | 1.0099 | 188.4 | 188.4 | 0.639 | 4.182 | 0.423 | 637.3 | 596.3 |
| 50.0 | 1.0121 | 209.3 | 209.3 | 0.704 | 4.182 | 0.457 | 643.5 | 547.1 |
| 60.0 | 1.0171 | 251.2 | 251.1 | 0.831 | 4.183 | 0.522 | 654.3 | 466.6 |
| 69.11 | | | | saturation | | | | |
| liquid | 1.0222 | 289.3 | 289.3 | 0.944 | 4.186 | 0.578 | 662.4 | 409.0 |
| vapour | 5229.8 | 2624 | 2467 | 7.767 | 1.956 | 3.024 | 21.99 | 11.23 |
| 70.0 | 5243.8 | 2626 | 2468 | 7.772 | 1.955 | 3.014 | 22.05 | 11.26 |
| 80.0 | 5401.5 | 2645 | 2483 | 7.828 | 1.945 | 2.914 | 22.75 | 11.61 |
| 90.0 | 5558.6 | 2665 | 2498 | 7.882 | 1.937 | 2.821 | 23.48 | 11.96 |
| 100.0 | 5715.2 | 2684 | 2513 | 7.935 | 1.933 | 2.736 | 24.24 | 12.33 |
| 125.0 | 6105.1 | 2732 | 2549 | 8.060 | 1.929 | 2.549 | 26.25 | 13.26 |
| 150.0 | 6493.5 | 2781 | 2586 | 8.178 | 1.933 | 2.389 | 28.39 | 14.22 |
| 175.0 | 6880.9 | 2829 | 2623 | 8.289 | 1.940 | 2.250 | 30.65 | 15.20 |
| 200.0 | 7267.6 | 2878 | 2660 | 8.394 | 1.950 | 2.127 | 33.01 | 16.20 |
| 225.0 | 7653.9 | 2927 | 2697 | 8.495 | 1.962 | 2.018 | 35.46 | 17.21 |
| 250.0 | 8039.8 | 2976 | 2735 | 8.592 | 1.975 | 1.919 | 38.00 | 18.23 |
| 275.0 | 8425.5 | 3025 | 2773 | 8.684 | 1.989 | 1.831 | 40.62 | 19.26 |
| 300.0 | 8811.0 | 3075 | 2811 | 8.773 | 2.004 | 1.750 | 43.30 | 20.30 |
| 325.0 | 9196.3 | 3125 | 2850 | 8.859 | 2.018 | 1.676 | 46.06 | 21.33 |
| 350.0 | 9581.6 | 3176 | 2889 | 8.942 | 2.034 | 1.608 | 48.88 | 22.37 |
| 375.0 | 9966.7 | 3227 | 2928 | 9.022 | 2.050 | 1.546 | 51.76 | 23.41 |
| 400.0 | 10351 | 3279 | 2968 | 9.100 | 2.066 | 1.488 | 54.69 | 24.45 |
| 450.0 | 11121 | 3383 | 3049 | 9.249 | 2.098 | 1.384 | 60.71 | 26.52 |
| 500.0 | 11891 | 3488 | 3132 | 9.391 | 2.132 | 1.295 | 66.92 | 28.57 |
| 550.0 | 12661 | 3596 | 3216 | 9.525 | 2.166 | 1.216 | 73.31 | 30.60 |
| 600.0 | 13430 | 3705 | 3302 | 9.654 | 2.201 | 1.146 | 79.85 | 32.61 |
| 650.0 | 14200 | 3816 | 3390 | 9.778 | 2.237 | 1.084 | 86.53 | 34.59 |
| 700.0 | 14969 | 3929 | 3480 | 9.896 | 2.272 | 1.028 | 93.33 | 36.55 |
| 750.0 | 15739 | 4043 | 3571 | 10.011 | 2.308 | 0.978 | 100.3 | 38.47 |
| 800.0 | 16508 | 4159 | 3664 | 10.122 | 2.343 | 0.932 | 107.3 | 40.37 |
| 850.0 | 17277 | 4278 | 3759 | 10.230 | 2.377 | 0.891 | 114.4 | 42.24 |
| 900.0 | 18047 | 4397 | 3856 | 10.334 | 2.412 | 0.853 | 121.6 | 44.07 |
| 950.0 | 18816 | 4519 | 3954 | 10.435 | 2.445 | 0.818 | 128.9 | 45.88 |
| 1000.0 | 19585 | 4642 | 4054 | 10.534 | 2.478 | 0.786 | 136.3 | 47.66 |

Table 1 (cont.) $p = 0.04\,\text{MPa} = 0.4\ \text{bar}$

| $t$ | $v$ | $h$ | $u$ | $s$ | $c_\text{p}$ | $\alpha_\text{p}$ | $\lambda$ | $\eta$ |
|---|---|---|---|---|---|---|---|---|
| °C | dm³/kg | kJ/kg | kJ/kg | kJ/kg K | kJ/kg K | $10^{-3}$/K | mW/K m | $\mu$Pa s |
| 0.0 | 1.0002 | −0.001 | −0.04 | −0.0002 | 4.229 | −0.080 | 561.0 | 1792 |
| 5.0 | 1.0000 | 21.1 | 21.0 | 0.076 | 4.200 | 0.011 | 570.5 | 1518 |
| 10.0 | 1.0003 | 42.0 | 42.0 | 0.151 | 4.188 | 0.087 | 580.0 | 1306 |
| 15.0 | 1.0009 | 63.0 | 62.9 | 0.224 | 4.184 | 0.152 | 589.3 | 1138 |
| 20.0 | 1.0018 | 83.9 | 83.8 | 0.296 | 4.183 | 0.209 | 598.4 | 1002 |
| 25.0 | 1.0030 | 104.8 | 104.7 | 0.367 | 4.183 | 0.259 | 607.1 | 890.5 |
| 30.0 | 1.0044 | 125.7 | 125.7 | 0.437 | 4.183 | 0.305 | 615.4 | 797.7 |
| 35.0 | 1.0060 | 146.6 | 146.6 | 0.505 | 4.183 | 0.347 | 623.3 | 719.6 |
| 40.0 | 1.0079 | 167.5 | 167.5 | 0.572 | 4.182 | 0.386 | 630.6 | 653.2 |
| 45.0 | 1.0099 | 188.4 | 188.4 | 0.639 | 4.182 | 0.423 | 637.3 | 596.3 |
| 50.0 | 1.0121 | 209.4 | 209.3 | 0.704 | 4.182 | 0.457 | 643.5 | 547.1 |
| 60.0 | 1.0171 | 251.2 | 251.1 | 0.831 | 4.183 | 0.522 | 654.3 | 466.6 |
| 70.0 | 1.0228 | 293.0 | 293.0 | 0.955 | 4.187 | 0.583 | 663.1 | 404.1 |
| 75.88 | | | | saturation | | | | |
| liquid | 1.0264 | 317.6 | 317.6 | 1.026 | 4.191 | 0.617 | 667.3 | 373.6 |
| vapour | 3994.0 | 2636 | 2476 | 7.669 | 1.972 | 2.985 | 22.61 | 11.45 |
| 80.0 | 4043.1 | 2644 | 2482 | 7.692 | 1.966 | 2.942 | 22.90 | 11.60 |
| 90.0 | 4161.8 | 2663 | 2497 | 7.747 | 1.955 | 2.845 | 23.61 | 11.96 |
| 100.0 | 4279.9 | 2683 | 2512 | 7.800 | 1.948 | 2.756 | 24.36 | 12.32 |
| 125.0 | 4573.7 | 2731 | 2548 | 7.926 | 1.939 | 2.562 | 26.33 | 13.25 |
| 150.0 | 4866.0 | 2780 | 2585 | 8.044 | 1.939 | 2.398 | 28.45 | 14.21 |
| 175.0 | 5157.3 | 2828 | 2622 | 8.155 | 1.945 | 2.256 | 30.70 | 15.19 |
| 200.0 | 5447.9 | 2877 | 2659 | 8.261 | 1.954 | 2.132 | 33.05 | 16.19 |
| 225.0 | 5738.0 | 2926 | 2697 | 8.362 | 1.965 | 2.021 | 35.49 | 17.21 |
| 250.0 | 6027.8 | 2975 | 2734 | 8.458 | 1.977 | 1.922 | 38.02 | 18.23 |
| 275.0 | 6317.3 | 3025 | 2772 | 8.551 | 1.991 | 1.833 | 40.64 | 19.26 |
| 300.0 | 6606.6 | 3075 | 2811 | 8.640 | 2.005 | 1.751 | 43.32 | 20.30 |
| 325.0 | 6895.8 | 3125 | 2849 | 8.726 | 2.020 | 1.677 | 46.07 | 21.33 |
| 350.0 | 7184.9 | 3176 | 2889 | 8.809 | 2.035 | 1.609 | 48.89 | 22.37 |
| 375.0 | 7473.9 | 3227 | 2928 | 8.889 | 2.050 | 1.546 | 51.77 | 23.41 |
| 400.0 | 7762.8 | 3278 | 2968 | 8.967 | 2.066 | 1.489 | 54.70 | 24.45 |
| 450.0 | 8340.5 | 3383 | 3049 | 9.116 | 2.099 | 1.385 | 60.72 | 26.52 |
| 500.0 | 8917.9 | 3488 | 3132 | 9.258 | 2.132 | 1.295 | 66.93 | 28.57 |
| 550.0 | 9495.3 | 3596 | 3216 | 9.392 | 2.167 | 1.216 | 73.31 | 30.60 |
| 600.0 | 10072 | 3705 | 3302 | 9.521 | 2.202 | 1.146 | 79.85 | 32.61 |
| 650.0 | 10649 | 3816 | 3390 | 9.645 | 2.237 | 1.084 | 86.53 | 34.60 |
| 700.0 | 11226 | 3929 | 3480 | 9.764 | 2.272 | 1.028 | 93.34 | 36.55 |
| 750.0 | 11804 | 4043 | 3571 | 9.878 | 2.308 | 0.978 | 100.3 | 38.48 |
| 800.0 | 12381 | 4159 | 3664 | 9.989 | 2.343 | 0.932 | 107.3 | 40.37 |
| 850.0 | 12958 | 4277 | 3759 | 10.097 | 2.378 | 0.891 | 114.4 | 42.24 |
| 900.0 | 13535 | 4397 | 3856 | 10.201 | 2.412 | 0.853 | 121.6 | 44.07 |
| 950.0 | 14112 | 4519 | 3954 | 10.302 | 2.445 | 0.818 | 128.9 | 45.88 |
| 1000.0 | 14689 | 4642 | 4054 | 10.401 | 2.478 | 0.786 | 136.3 | 47.66 |

Table 1 (cont.)

$$p = 0.05\,\text{MPa} = 0.5\ \text{bar}$$

| $t$ | $v$ | $h$ | $u$ | $s$ | $c_\mathrm{p}$ | $\alpha_\mathrm{p}$ | $\lambda$ | $\eta$ |
|---|---|---|---|---|---|---|---|---|
| °C | dm³/kg | kJ/kg | kJ/kg | kJ/kg K | kJ/kg K | $10^{-3}$/K | mW/K m | $\mu$Pa s |
| 0.0 | 1.0002 | 0.009 | −0.04 | −0.0002 | 4.228 | −0.080 | 561.0 | 1792 |
| 5.0 | 1.0000 | 21.1 | 21.0 | 0.076 | 4.200 | 0.011 | 570.5 | 1518 |
| 10.0 | 1.0003 | 42.0 | 42.0 | 0.151 | 4.188 | 0.087 | 580.0 | 1306 |
| 15.0 | 1.0009 | 63.0 | 62.9 | 0.224 | 4.184 | 0.152 | 589.4 | 1137 |
| 20.0 | 1.0018 | 83.9 | 83.8 | 0.296 | 4.183 | 0.209 | 598.4 | 1002 |
| 25.0 | 1.0030 | 104.8 | 104.7 | 0.367 | 4.183 | 0.259 | 607.1 | 890.5 |
| 30.0 | 1.0044 | 125.7 | 125.7 | 0.437 | 4.183 | 0.305 | 615.4 | 797.7 |
| 35.0 | 1.0060 | 146.6 | 146.6 | 0.505 | 4.183 | 0.347 | 623.3 | 719.6 |
| 40.0 | 1.0079 | 167.5 | 167.5 | 0.572 | 4.182 | 0.386 | 630.6 | 653.2 |
| 45.0 | 1.0099 | 188.5 | 188.4 | 0.639 | 4.182 | 0.423 | 637.3 | 596.3 |
| 50.0 | 1.0121 | 209.4 | 209.3 | 0.704 | 4.182 | 0.457 | 643.5 | 547.1 |
| 60.0 | 1.0171 | 251.2 | 251.1 | 0.831 | 4.183 | 0.522 | 654.3 | 466.6 |
| 70.0 | 1.0227 | 293.0 | 293.0 | 0.955 | 4.187 | 0.583 | 663.1 | 404.1 |
| 80.0 | 1.0290 | 334.9 | 334.9 | 1.075 | 4.194 | 0.640 | 670.0 | 354.5 |
| 81.34 | saturation | | | | | | | |
| liquid | 1.0299 | 340.5 | 340.5 | 1.091 | 4.195 | 0.648 | 670.8 | 348.6 |
| vapour | 3240.9 | 2645 | 2483 | 7.593 | 1.986 | 2.957 | 23.14 | 11.64 |
| 90.0 | 3323.6 | 2662 | 2496 | 7.641 | 1.973 | 2.869 | 23.74 | 11.95 |
| 100.0 | 3418.8 | 2682 | 2511 | 7.694 | 1.963 | 2.776 | 24.47 | 12.31 |
| 125.0 | 3654.9 | 2730 | 2548 | 7.821 | 1.949 | 2.575 | 26.42 | 13.24 |
| 150.0 | 3889.5 | 2779 | 2585 | 7.939 | 1.946 | 2.407 | 28.52 | 14.21 |
| 175.0 | 4123.1 | 2828 | 2622 | 8.051 | 1.950 | 2.262 | 30.75 | 15.19 |
| 200.0 | 4356.0 | 2877 | 2659 | 8.157 | 1.957 | 2.136 | 33.08 | 16.19 |
| 225.0 | 4588.4 | 2926 | 2696 | 8.258 | 1.968 | 2.025 | 35.52 | 17.20 |
| 250.0 | 4820.5 | 2975 | 2734 | 8.355 | 1.979 | 1.925 | 38.05 | 18.23 |
| 275.0 | 5052.4 | 3025 | 2772 | 8.448 | 1.992 | 1.835 | 40.66 | 19.26 |
| 300.0 | 5284.0 | 3075 | 2810 | 8.537 | 2.006 | 1.753 | 43.34 | 20.29 |
| 325.0 | 5515.5 | 3125 | 2849 | 8.623 | 2.021 | 1.678 | 46.09 | 21.33 |
| 350.0 | 5746.9 | 3176 | 2888 | 8.706 | 2.036 | 1.610 | 48.90 | 22.37 |
| 375.0 | 5978.2 | 3227 | 2928 | 8.786 | 2.051 | 1.547 | 51.78 | 23.41 |
| 400.0 | 6209.4 | 3278 | 2968 | 8.864 | 2.067 | 1.489 | 54.71 | 24.45 |
| 450.0 | 6671.7 | 3382 | 3049 | 9.013 | 2.099 | 1.385 | 60.73 | 26.52 |
| 500.0 | 7133.8 | 3488 | 3132 | 9.155 | 2.133 | 1.295 | 66.94 | 28.57 |
| 550.0 | 7595.8 | 3596 | 3216 | 9.289 | 2.167 | 1.216 | 73.32 | 30.60 |
| 600.0 | 8057.7 | 3705 | 3302 | 9.418 | 2.202 | 1.146 | 79.86 | 32.61 |
| 650.0 | 8519.5 | 3816 | 3390 | 9.542 | 2.237 | 1.084 | 86.54 | 34.60 |
| 700.0 | 8981.3 | 3929 | 3480 | 9.661 | 2.272 | 1.028 | 93.35 | 36.55 |
| 750.0 | 9443.0 | 4043 | 3571 | 9.775 | 2.308 | 0.978 | 100.3 | 38.48 |
| 800.0 | 9904.7 | 4159 | 3664 | 9.886 | 2.343 | 0.932 | 107.3 | 40.37 |
| 850.0 | 10366 | 4277 | 3759 | 9.994 | 2.378 | 0.891 | 114.4 | 42.24 |
| 900.0 | 10828 | 4397 | 3856 | 10.098 | 2.412 | 0.853 | 121.6 | 44.07 |
| 950.0 | 11289 | 4519 | 3954 | 10.199 | 2.445 | 0.818 | 128.9 | 45.88 |
| 1000.0 | 11751 | 4642 | 4054 | 10.298 | 2.478 | 0.786 | 136.3 | 47.66 |

Table 1 (cont.)

$$p = 0.075\,\text{MPa} = 0.75\ \text{bar}$$

| $t$ | $v$ | $h$ | $u$ | $s$ | $c_p$ | $\alpha_p$ | $\lambda$ | $\eta$ |
|---|---|---|---|---|---|---|---|---|
| °C | dm³/kg | kJ/kg | kJ/kg | kJ/kg K | kJ/kg K | $10^{-3}$/K | mW/K m | µPa s |
| 0.0 | 1.0002 | 0.03 | −0.04 | −0.0001 | 4.228 | −0.080 | 561.0 | 1792 |
| 5.0 | 1.0000 | 21.1 | 21.0 | 0.076 | 4.200 | 0.011 | 570.6 | 1518 |
| 10.0 | 1.0003 | 42.1 | 42.0 | 0.151 | 4.188 | 0.087 | 580.0 | 1306 |
| 15.0 | 1.0009 | 63.0 | 62.9 | 0.224 | 4.184 | 0.152 | 589.4 | 1137 |
| 20.0 | 1.0018 | 83.9 | 83.8 | 0.296 | 4.183 | 0.209 | 598.4 | 1002 |
| 25.0 | 1.0030 | 104.8 | 104.7 | 0.367 | 4.183 | 0.259 | 607.2 | 890.5 |
| 30.0 | 1.0044 | 125.7 | 125.7 | 0.437 | 4.183 | 0.305 | 615.5 | 797.7 |
| 35.0 | 1.0060 | 146.7 | 146.6 | 0.505 | 4.183 | 0.347 | 623.3 | 719.6 |
| 40.0 | 1.0079 | 167.6 | 167.5 | 0.572 | 4.182 | 0.386 | 630.6 | 653.2 |
| 45.0 | 1.0099 | 188.5 | 188.4 | 0.639 | 4.182 | 0.423 | 637.3 | 596.3 |
| 50.0 | 1.0121 | 209.4 | 209.3 | 0.704 | 4.182 | 0.457 | 643.5 | 547.1 |
| 60.0 | 1.0171 | 251.2 | 251.1 | 0.831 | 4.183 | 0.522 | 654.3 | 466.6 |
| 70.0 | 1.0227 | 293.0 | 293.0 | 0.955 | 4.187 | 0.583 | 663.1 | 404.1 |
| 80.0 | 1.0290 | 334.9 | 334.9 | 1.075 | 4.194 | 0.640 | 670.0 | 354.5 |
| 90.0 | 1.0359 | 376.9 | 376.9 | 1.193 | 4.204 | 0.696 | 675.3 | 314.5 |
| 91.78 | | | | saturation | | | | |
| liquid | 1.0372 | 384.4 | 384.4 | 1.213 | 4.206 | 0.706 | 676.0 | 308.2 |
| vapour | 2217.5 | 2662 | 2496 | 7.456 | 2.017 | 2.912 | 24.20 | 11.99 |
| 100.0 | 2270.4 | 2679 | 2508 | 7.500 | 2.001 | 2.827 | 24.77 | 12.29 |
| 125.0 | 2429.8 | 2728 | 2546 | 7.629 | 1.974 | 2.608 | 26.64 | 13.23 |
| 150.0 | 2587.5 | 2777 | 2583 | 7.749 | 1.963 | 2.429 | 28.69 | 14.19 |
| 175.0 | 2744.2 | 2826 | 2621 | 7.862 | 1.962 | 2.278 | 30.87 | 15.18 |
| 200.0 | 2900.2 | 2875 | 2658 | 7.968 | 1.966 | 2.148 | 33.18 | 16.18 |
| 225.0 | 3055.7 | 2925 | 2696 | 8.070 | 1.974 | 2.033 | 35.60 | 17.20 |
| 250.0 | 3210.9 | 2974 | 2733 | 8.167 | 1.985 | 1.931 | 38.11 | 18.22 |
| 275.0 | 3365.8 | 3024 | 2772 | 8.259 | 1.997 | 1.840 | 40.70 | 19.26 |
| 300.0 | 3520.5 | 3074 | 2810 | 8.349 | 2.010 | 1.757 | 43.38 | 20.29 |
| 325.0 | 3675.1 | 3124 | 2849 | 8.435 | 2.023 | 1.682 | 46.12 | 21.33 |
| 350.0 | 3829.6 | 3175 | 2888 | 8.518 | 2.038 | 1.613 | 48.93 | 22.37 |
| 375.0 | 3984.0 | 3226 | 2928 | 8.598 | 2.053 | 1.550 | 51.80 | 23.41 |
| 400.0 | 4138.3 | 3278 | 2968 | 8.676 | 2.068 | 1.491 | 54.73 | 24.45 |
| 450.0 | 4446.7 | 3382 | 3049 | 8.826 | 2.100 | 1.387 | 60.75 | 26.52 |
| 500.0 | 4755.0 | 3488 | 3131 | 8.967 | 2.134 | 1.296 | 66.96 | 28.57 |
| 550.0 | 5063.1 | 3595 | 3216 | 9.102 | 2.168 | 1.217 | 73.34 | 30.61 |
| 600.0 | 5371.2 | 3705 | 3302 | 9.231 | 2.202 | 1.147 | 79.87 | 32.61 |
| 650.0 | 5679.1 | 3816 | 3390 | 9.354 | 2.237 | 1.085 | 86.55 | 34.60 |
| 700.0 | 5987.1 | 3928 | 3479 | 9.473 | 2.273 | 1.029 | 93.36 | 36.55 |
| 750.0 | 6295.0 | 4043 | 3571 | 9.588 | 2.308 | 0.978 | 100.3 | 38.48 |
| 800.0 | 6602.8 | 4159 | 3664 | 9.699 | 2.343 | 0.932 | 107.3 | 40.37 |
| 850.0 | 6910.7 | 4277 | 3759 | 9.807 | 2.378 | 0.891 | 114.4 | 42.24 |
| 900.0 | 7218.5 | 4397 | 3856 | 9.911 | 2.412 | 0.853 | 121.7 | 44.08 |
| 950.0 | 7526.3 | 4519 | 3954 | 10.012 | 2.445 | 0.818 | 128.9 | 45.88 |
| 1000.0 | 7834.1 | 4642 | 4054 | 10.111 | 2.478 | 0.786 | 136.3 | 47.66 |

Table 1 (cont.)

$$p = 0.10\,\mathrm{MPa} = 1.0\ \mathrm{bar}$$

| $t$ | $v$ | $h$ | $u$ | $s$ | $c_\mathrm{p}$ | $\alpha_\mathrm{p}$ | $\lambda$ | $\eta$ |
|---|---|---|---|---|---|---|---|---|
| °C | dm³/kg | kJ/kg | kJ/kg | kJ/kg K | kJ/kg K | $10^{-3}$/K | mW/K m | µPa s |
| 0.0 | 1.0002 | 0.06 | −0.04 | −0.0001 | 4.228 | −0.080 | 561.0 | 1792 |
| 5.0 | 1.0000 | 21.1 | 21.0 | 0.076 | 4.200 | 0.011 | 570.6 | 1518 |
| 10.0 | 1.0003 | 42.1 | 42.0 | 0.151 | 4.188 | 0.087 | 580.0 | 1306 |
| 15.0 | 1.0009 | 63.0 | 62.9 | 0.224 | 4.184 | 0.152 | 589.4 | 1137 |
| 20.0 | 1.0018 | 83.9 | 83.8 | 0.296 | 4.183 | 0.209 | 598.4 | 1001 |
| 25.0 | 1.0029 | 104.8 | 104.7 | 0.367 | 4.183 | 0.259 | 607.2 | 890.4 |
| 30.0 | 1.0044 | 125.8 | 125.7 | 0.437 | 4.183 | 0.305 | 615.5 | 797.7 |
| 35.0 | 1.0060 | 146.7 | 146.6 | 0.505 | 4.183 | 0.347 | 623.3 | 719.6 |
| 40.0 | 1.0079 | 167.6 | 167.5 | 0.572 | 4.182 | 0.386 | 630.6 | 653.3 |
| 45.0 | 1.0099 | 188.5 | 188.4 | 0.638 | 4.182 | 0.423 | 637.3 | 596.3 |
| 50.0 | 1.0121 | 209.4 | 209.3 | 0.704 | 4.181 | 0.457 | 643.6 | 547.1 |
| 60.0 | 1.0171 | 251.2 | 251.1 | 0.831 | 4.183 | 0.522 | 654.4 | 466.6 |
| 70.0 | 1.0227 | 293.1 | 293.0 | 0.955 | 4.187 | 0.583 | 663.1 | 404.1 |
| 80.0 | 1.0290 | 335.0 | 334.9 | 1.075 | 4.194 | 0.640 | 670.0 | 354.5 |
| 90.0 | 1.0359 | 377.0 | 376.9 | 1.193 | 4.204 | 0.696 | 675.3 | 314.6 |
| 99.63 | | | | saturation | | | | |
| liquid | 1.0431 | 417.5 | 417.4 | 1.303 | 4.217 | 0.748 | 679.0 | 283.0 |
| vapour | 1694.3 | 2675 | 2505 | 7.359 | 2.043 | 2.885 | 25.05 | 12.26 |
| 100.0 | 1696.1 | 2675 | 2506 | 7.361 | 2.042 | 2.881 | 25.08 | 12.27 |
| 125.0 | 1817.1 | 2726 | 2544 | 7.492 | 1.999 | 2.642 | 26.87 | 13.21 |
| 150.0 | 1936.4 | 2776 | 2582 | 7.613 | 1.980 | 2.452 | 28.85 | 14.18 |
| 175.0 | 2054.7 | 2825 | 2619 | 7.726 | 1.974 | 2.295 | 31.00 | 15.17 |
| 200.0 | 2172.3 | 2874 | 2657 | 7.833 | 1.975 | 2.160 | 33.28 | 16.18 |
| 225.0 | 2289.3 | 2924 | 2695 | 7.935 | 1.981 | 2.042 | 35.68 | 17.19 |
| 250.0 | 2406.1 | 2973 | 2733 | 8.033 | 1.990 | 1.938 | 38.17 | 18.22 |
| 275.0 | 2522.5 | 3023 | 2771 | 8.126 | 2.001 | 1.845 | 40.75 | 19.25 |
| 300.0 | 2638.8 | 3073 | 2810 | 8.215 | 2.013 | 1.761 | 43.42 | 20.29 |
| 325.0 | 2754.9 | 3124 | 2848 | 8.301 | 2.026 | 1.685 | 46.16 | 21.33 |
| 350.0 | 2870.9 | 3175 | 2888 | 8.385 | 2.040 | 1.616 | 48.96 | 22.37 |
| 375.0 | 2986.9 | 3226 | 2927 | 8.465 | 2.055 | 1.552 | 51.83 | 23.41 |
| 400.0 | 3102.7 | 3278 | 2967 | 8.543 | 2.070 | 1.493 | 54.76 | 24.45 |
| 450.0 | 3334.2 | 3382 | 3048 | 8.693 | 2.102 | 1.388 | 60.77 | 26.52 |
| 500.0 | 3565.5 | 3488 | 3131 | 8.834 | 2.135 | 1.297 | 66.97 | 28.57 |
| 550.0 | 3796.8 | 3595 | 3216 | 8.969 | 2.168 | 1.218 | 73.35 | 30.61 |
| 600.0 | 4027.9 | 3705 | 3302 | 9.098 | 2.203 | 1.147 | 79.89 | 32.61 |
| 650.0 | 4259.0 | 3816 | 3390 | 9.222 | 2.238 | 1.085 | 86.57 | 34.60 |
| 700.0 | 4490.0 | 3928 | 3479 | 9.341 | 2.273 | 1.029 | 93.37 | 36.55 |
| 750.0 | 4721.0 | 4043 | 3571 | 9.455 | 2.308 | 0.978 | 100.3 | 38.48 |
| 800.0 | 4951.9 | 4159 | 3664 | 9.566 | 2.343 | 0.933 | 107.3 | 40.37 |
| 850.0 | 5182.8 | 4277 | 3759 | 9.674 | 2.378 | 0.891 | 114.4 | 42.24 |
| 900.0 | 5413.7 | 4397 | 3856 | 9.778 | 2.412 | 0.853 | 121.7 | 44.08 |
| 950.0 | 5644.6 | 4518 | 3954 | 9.879 | 2.446 | 0.818 | 129.0 | 45.88 |
| 1000.0 | 5875.5 | 4642 | 4054 | 9.978 | 2.478 | 0.786 | 136.3 | 47.66 |

Table 1 (cont.)

$$p = 0.50\,\text{MPa} = 5.0\ \text{bar}$$

| $t$ | $v$ | $h$ | $u$ | $s$ | $c_p$ | $\alpha_p$ | $\lambda$ | $\eta$ |
|---|---|---|---|---|---|---|---|---|
| °C | dm³/kg | kJ/kg | kJ/kg | kJ/kg K | kJ/kg K | $10^{-3}$/K | mW/K m | $\mu$Pa s |
| 0.0 | 1.0000 | 0.5 | −0.03 | −0.0001 | 4.226 | −0.079 | 561.2 | 1791 |
| 5.0 | 0.9998 | 21.5 | 21.0 | 0.076 | 4.198 | 0.013 | 570.8 | 1517 |
| 10.0 | 1.0001 | 42.5 | 42.0 | 0.151 | 4.186 | 0.088 | 580.2 | 1305 |
| 15.0 | 1.0007 | 63.4 | 62.9 | 0.224 | 4.182 | 0.153 | 589.6 | 1137 |
| 20.0 | 1.0016 | 84.3 | 83.8 | 0.296 | 4.182 | 0.210 | 598.6 | 1001 |
| 25.0 | 1.0028 | 105.2 | 104.7 | 0.367 | 4.182 | 0.260 | 607.3 | 890.4 |
| 30.0 | 1.0042 | 126.1 | 125.6 | 0.436 | 4.182 | 0.305 | 615.6 | 797.6 |
| 35.0 | 1.0058 | 147.0 | 146.5 | 0.505 | 4.182 | 0.347 | 623.5 | 719.6 |
| 40.0 | 1.0077 | 167.9 | 167.4 | 0.572 | 4.181 | 0.386 | 630.8 | 653.3 |
| 45.0 | 1.0097 | 188.8 | 188.3 | 0.638 | 4.181 | 0.423 | 637.5 | 596.4 |
| 50.0 | 1.0119 | 209.7 | 209.2 | 0.704 | 4.181 | 0.457 | 643.7 | 547.1 |
| 60.0 | 1.0169 | 251.6 | 251.0 | 0.831 | 4.182 | 0.522 | 654.5 | 466.7 |
| 70.0 | 1.0225 | 293.4 | 292.9 | 0.955 | 4.186 | 0.582 | 663.3 | 404.2 |
| 80.0 | 1.0288 | 335.3 | 334.8 | 1.075 | 4.193 | 0.640 | 670.2 | 354.6 |
| 90.0 | 1.0357 | 377.3 | 376.7 | 1.192 | 4.203 | 0.695 | 675.5 | 314.7 |
| 100.0 | 1.0432 | 419.4 | 418.8 | 1.307 | 4.216 | 0.749 | 679.3 | 282.0 |
| 125.0 | 1.0647 | 525.3 | 524.7 | 1.581 | 4.257 | 0.884 | 683.8 | 222.3 |
| 150.0 | 1.0904 | 632.3 | 631.8 | 1.842 | 4.312 | 1.026 | 682.1 | 182.5 |
| 151.87 | | | | saturation | | | | |
| liquid | 1.0925 | 640.4 | 639.8 | 1.861 | 4.317 | 1.038 | 681.7 | 180.1 |
| vapour | 374.86 | 2748 | 2561 | 6.821 | 2.312 | 2.884 | 31.87 | 14.06 |
| 175.0 | 399.33 | 2800 | 2601 | 6.941 | 2.204 | 2.599 | 33.17 | 15.01 |
| 200.0 | 424.87 | 2854 | 2642 | 7.059 | 2.138 | 2.372 | 34.93 | 16.05 |
| 225.0 | 449.80 | 2907 | 2682 | 7.168 | 2.101 | 2.196 | 36.95 | 17.09 |
| 250.0 | 474.32 | 2960 | 2722 | 7.270 | 2.081 | 2.054 | 39.18 | 18.14 |
| 275.0 | 498.54 | 3011 | 2762 | 7.367 | 2.072 | 1.934 | 41.57 | 19.19 |
| 300.0 | 522.55 | 3063 | 2802 | 7.459 | 2.069 | 1.831 | 44.09 | 20.24 |
| 325.0 | 546.40 | 3115 | 2842 | 7.547 | 2.072 | 1.741 | 46.72 | 21.29 |
| 350.0 | 570.12 | 3167 | 2882 | 7.632 | 2.078 | 1.661 | 49.44 | 22.34 |
| 375.0 | 593.74 | 3219 | 2922 | 7.714 | 2.087 | 1.589 | 52.25 | 23.39 |
| 400.0 | 617.29 | 3271 | 2963 | 7.794 | 2.097 | 1.524 | 55.13 | 24.44 |
| 450.0 | 664.20 | 3377 | 3045 | 7.945 | 2.121 | 1.410 | 61.08 | 26.52 |
| 500.0 | 710.94 | 3483 | 3128 | 8.087 | 2.149 | 1.313 | 67.25 | 28.58 |
| 550.0 | 757.55 | 3592 | 3213 | 8.223 | 2.180 | 1.229 | 73.61 | 30.62 |
| 600.0 | 804.08 | 3701 | 3299 | 8.352 | 2.212 | 1.156 | 80.13 | 32.63 |
| 650.0 | 850.53 | 3813 | 3388 | 8.477 | 2.245 | 1.092 | 86.80 | 34.61 |
| 700.0 | 896.94 | 3926 | 3478 | 8.596 | 2.279 | 1.034 | 93.59 | 36.57 |
| 750.0 | 943.31 | 4041 | 3569 | 8.711 | 2.313 | 0.983 | 100.5 | 38.49 |
| 800.0 | 989.65 | 4157 | 3663 | 8.822 | 2.348 | 0.936 | 107.5 | 40.39 |
| 850.0 | 1036.0 | 4276 | 3758 | 8.930 | 2.382 | 0.894 | 114.6 | 42.26 |
| 900.0 | 1082.2 | 4396 | 3854 | 9.034 | 2.415 | 0.855 | 121.8 | 44.09 |
| 950.0 | 1128.5 | 4517 | 3953 | 9.136 | 2.448 | 0.820 | 129.1 | 45.90 |
| 1000.0 | 1174.8 | 4640 | 4053 | 9.234 | 2.481 | 0.787 | 136.4 | 47.68 |

Table 1 (cont.)

$$p = 1.0 \, \text{MPa} = 10 \; \text{bar}$$

| $t$ | $v$ | $h$ | $u$ | $s$ | $c_p$ | $\alpha_p$ | $\lambda$ | $\eta$ |
|---|---|---|---|---|---|---|---|---|
| °C | dm³/kg | kJ/kg | kJ/kg | kJ/kg K | kJ/kg K | $10^{-3}$/K | mW/K m | μPa s |
| 0.0 | 0.9997 | 1. | −0.02 | −0.00008 | 4.223 | −0.076 | 561.5 | 1790 |
| 5.0 | 0.9996 | 22.0 | 21.0 | 0.076 | 4.196 | 0.014 | 571.0 | 1517 |
| 10.0 | 0.9998 | 43.0 | 42.0 | 0.151 | 4.184 | 0.090 | 580.5 | 1305 |
| 15.0 | 1.0004 | 63.9 | 62.9 | 0.224 | 4.181 | 0.154 | 589.8 | 1137 |
| 20.0 | 1.0014 | 84.8 | 83.8 | 0.296 | 4.180 | 0.210 | 598.9 | 1001 |
| 25.0 | 1.0025 | 105.7 | 104.7 | 0.367 | 4.181 | 0.260 | 607.6 | 890.2 |
| 30.0 | 1.0040 | 126.6 | 125.6 | 0.436 | 4.181 | 0.306 | 615.9 | 797.6 |
| 35.0 | 1.0056 | 147.5 | 146.5 | 0.505 | 4.180 | 0.347 | 623.7 | 719.6 |
| 40.0 | 1.0075 | 168.4 | 167.4 | 0.572 | 4.180 | 0.386 | 631.0 | 653.3 |
| 45.0 | 1.0095 | 189.3 | 188.3 | 0.638 | 4.180 | 0.423 | 637.8 | 596.4 |
| 50.0 | 1.0117 | 210.2 | 209.2 | 0.703 | 4.179 | 0.457 | 644.0 | 547.2 |
| 60.0 | 1.0167 | 252.0 | 251.0 | 0.831 | 4.181 | 0.522 | 654.8 | 466.8 |
| 70.0 | 1.0223 | 293.8 | 292.8 | 0.954 | 4.185 | 0.582 | 663.6 | 404.3 |
| 80.0 | 1.0286 | 335.7 | 334.7 | 1.075 | 4.192 | 0.639 | 670.5 | 354.7 |
| 90.0 | 1.0355 | 377.7 | 376.6 | 1.192 | 4.202 | 0.694 | 675.8 | 314.8 |
| 100.0 | 1.0430 | 419.7 | 418.7 | 1.306 | 4.215 | 0.749 | 679.6 | 282.1 |
| 125.0 | 1.0644 | 525.6 | 524.5 | 1.581 | 4.256 | 0.883 | 684.1 | 222.4 |
| 150.0 | 1.0901 | 632.6 | 631.5 | 1.841 | 4.310 | 1.024 | 682.4 | 182.7 |
| 175.0 | 1.1206 | 741.3 | 740.2 | 2.091 | 4.385 | 1.185 | 675.4 | 154.8 |
| 179.92 | saturation | | | | | | | |
| liquid | 1.1272 | 762.9 | 761.8 | 2.139 | 4.403 | 1.220 | 673.4 | 150.3 |
| vapour | 194.38 | 2777 | 2583 | 6.586 | 2.557 | 3.038 | 36.43 | 15.02 |
| 200.0 | 205.90 | 2827 | 2621 | 6.693 | 2.400 | 2.714 | 37.21 | 15.89 |
| 225.0 | 219.53 | 2885 | 2666 | 6.814 | 2.282 | 2.430 | 38.67 | 16.97 |
| 250.0 | 232.64 | 2941 | 2709 | 6.923 | 2.213 | 2.221 | 40.51 | 18.04 |
| 275.0 | 245.41 | 2996 | 2751 | 7.026 | 2.171 | 2.058 | 42.63 | 19.11 |
| 300.0 | 257.93 | 3050 | 2792 | 7.122 | 2.147 | 1.926 | 44.95 | 20.18 |
| 325.0 | 270.27 | 3104 | 2833 | 7.213 | 2.134 | 1.815 | 47.44 | 21.25 |
| 350.0 | 282.47 | 3157 | 2874 | 7.301 | 2.128 | 1.720 | 50.06 | 22.31 |
| 375.0 | 294.57 | 3210 | 2915 | 7.384 | 2.128 | 1.637 | 52.79 | 23.37 |
| 400.0 | 306.58 | 3263 | 2957 | 7.465 | 2.132 | 1.563 | 55.61 | 24.42 |
| 450.0 | 330.43 | 3370 | 3040 | 7.618 | 2.147 | 1.437 | 61.48 | 26.51 |
| 500.0 | 354.10 | 3478 | 3124 | 7.762 | 2.168 | 1.333 | 67.60 | 28.58 |
| 550.0 | 377.64 | 3587 | 3209 | 7.899 | 2.195 | 1.244 | 73.93 | 30.63 |
| 600.0 | 401.09 | 3698 | 3296 | 8.029 | 2.224 | 1.168 | 80.43 | 32.64 |
| 650.0 | 424.48 | 3810 | 3385 | 8.154 | 2.255 | 1.100 | 87.09 | 34.63 |
| 700.0 | 447.81 | 3923 | 3475 | 8.274 | 2.287 | 1.041 | 93.87 | 36.59 |
| 750.0 | 471.10 | 4038 | 3567 | 8.389 | 2.320 | 0.988 | 100.8 | 38.52 |
| 800.0 | 494.37 | 4155 | 3661 | 8.500 | 2.353 | 0.941 | 107.8 | 40.41 |
| 850.0 | 517.60 | 4274 | 3756 | 8.608 | 2.386 | 0.897 | 114.8 | 42.28 |
| 900.0 | 540.82 | 4394 | 3853 | 8.713 | 2.419 | 0.858 | 122.0 | 44.11 |
| 950.0 | 564.01 | 4515 | 3951 | 8.815 | 2.452 | 0.822 | 129.2 | 45.92 |
| 1000.0 | 587.20 | 4639 | 4052 | 8.914 | 2.484 | 0.789 | 136.5 | 47.69 |

Table 1 (cont.)

$$p = 1.5\,\text{MPa} = 15\ \text{bar}$$

| $t$ | $v$ | $h$ | $u$ | $s$ | $c_\text{P}$ | $\alpha_\text{P}$ | $\lambda$ | $\eta$ |
|---|---|---|---|---|---|---|---|---|
| °C | dm³/kg | kJ/kg | kJ/kg | kJ/kg K | kJ/kg K | $10^{-3}$/K | mW/K m | $\mu$Pa s |
| 0.0 | 0.9995 | 1.5 | −0.01 | −0.00004 | 4.221 | −0.074 | 561.8 | 1788 |
| 5.0 | 0.9993 | 22.5 | 21.0 | 0.076 | 4.193 | 0.016 | 571.3 | 1516 |
| 10.0 | 0.9996 | 43.4 | 41.9 | 0.151 | 4.182 | 0.091 | 580.7 | 1304 |
| 15.0 | 1.0002 | 64.3 | 62.8 | 0.224 | 4.179 | 0.155 | 590.0 | 1136 |
| 20.0 | 1.0011 | 85.2 | 83.7 | 0.296 | 4.179 | 0.211 | 599.1 | 1001 |
| 25.0 | 1.0023 | 106.1 | 104.6 | 0.367 | 4.179 | 0.261 | 607.8 | 890.1 |
| 30.0 | 1.0037 | 127.0 | 125.5 | 0.436 | 4.179 | 0.306 | 616.1 | 797.5 |
| 35.0 | 1.0054 | 147.9 | 146.4 | 0.504 | 4.179 | 0.348 | 623.9 | 719.6 |
| 40.0 | 1.0072 | 168.8 | 167.3 | 0.572 | 4.179 | 0.386 | 631.2 | 653.4 |
| 45.0 | 1.0093 | 189.7 | 188.2 | 0.638 | 4.178 | 0.423 | 638.0 | 596.5 |
| 50.0 | 1.0115 | 210.6 | 209.1 | 0.703 | 4.178 | 0.457 | 644.2 | 547.3 |
| 60.0 | 1.0165 | 252.4 | 250.9 | 0.830 | 4.180 | 0.521 | 655.0 | 466.9 |
| 70.0 | 1.0221 | 294.2 | 292.7 | 0.954 | 4.184 | 0.582 | 663.8 | 404.4 |
| 80.0 | 1.0283 | 336.1 | 334.5 | 1.074 | 4.191 | 0.639 | 670.7 | 354.9 |
| 90.0 | 1.0352 | 378.0 | 376.5 | 1.192 | 4.201 | 0.694 | 676.0 | 314.9 |
| 100.0 | 1.0427 | 420.1 | 418.6 | 1.306 | 4.214 | 0.748 | 679.9 | 282.3 |
| 125.0 | 1.0641 | 525.9 | 524.3 | 1.580 | 4.254 | 0.882 | 684.4 | 222.5 |
| 150.0 | 1.0898 | 632.9 | 631.3 | 1.841 | 4.309 | 1.023 | 682.8 | 182.8 |
| 175.0 | 1.1201 | 741.5 | 739.9 | 2.090 | 4.383 | 1.183 | 675.8 | 154.9 |
| 198.33 | | | | saturation | | | | |
| liquid | 1.1538 | 844.9 | 843.1 | 2.315 | 4.481 | 1.363 | 664.4 | 135.6 |
| vapour | 131.72 | 2791 | 2593 | 6.444 | 2.775 | 3.218 | 39.78 | 15.66 |
| 200.0 | 132.43 | 2796 | 2597 | 6.454 | 2.752 | 3.179 | 39.80 | 15.73 |
| 225.0 | 142.50 | 2861 | 2647 | 6.588 | 2.509 | 2.721 | 40.56 | 16.85 |
| 250.0 | 151.92 | 2922 | 2694 | 6.708 | 2.369 | 2.418 | 41.94 | 17.95 |
| 275.0 | 160.92 | 2980 | 2739 | 6.816 | 2.285 | 2.199 | 43.75 | 19.04 |
| 300.0 | 169.65 | 3036 | 2782 | 6.917 | 2.233 | 2.031 | 45.86 | 20.13 |
| 325.0 | 178.17 | 3092 | 2825 | 7.011 | 2.201 | 1.896 | 48.20 | 21.21 |
| 350.0 | 186.55 | 3147 | 2867 | 7.101 | 2.182 | 1.784 | 50.70 | 22.28 |
| 375.0 | 194.82 | 3201 | 2909 | 7.187 | 2.172 | 1.688 | 53.35 | 23.35 |
| 400.0 | 203.00 | 3255 | 2951 | 7.269 | 2.168 | 1.605 | 56.11 | 24.41 |
| 450.0 | 219.17 | 3364 | 3035 | 7.424 | 2.172 | 1.466 | 61.89 | 26.51 |
| 500.0 | 235.15 | 3473 | 3120 | 7.570 | 2.188 | 1.353 | 67.96 | 28.59 |
| 550.0 | 251.00 | 3583 | 3206 | 7.708 | 2.210 | 1.259 | 74.26 | 30.64 |
| 600.0 | 266.76 | 3694 | 3294 | 7.839 | 2.235 | 1.179 | 80.75 | 32.66 |
| 650.0 | 282.46 | 3806 | 3382 | 7.964 | 2.264 | 1.109 | 87.39 | 34.65 |
| 700.0 | 298.10 | 3920 | 3473 | 8.084 | 2.295 | 1.048 | 94.16 | 36.61 |
| 750.0 | 313.70 | 4036 | 3565 | 8.200 | 2.326 | 0.994 | 101.0 | 38.54 |
| 800.0 | 329.27 | 4153 | 3659 | 8.312 | 2.359 | 0.945 | 108.0 | 40.43 |
| 850.0 | 344.82 | 4271 | 3754 | 8.420 | 2.391 | 0.901 | 115.1 | 42.30 |
| 900.0 | 360.34 | 4392 | 3851 | 8.525 | 2.423 | 0.861 | 122.2 | 44.13 |
| 950.0 | 375.85 | 4514 | 3950 | 8.626 | 2.455 | 0.825 | 129.4 | 45.94 |
| 1000.0 | 391.34 | 4637 | 4050 | 8.725 | 2.487 | 0.791 | 136.7 | 47.71 |

Table 1 (cont.) $p = 2.0\,\mathrm{MPa} = 20$ bar

| $t$ | $v$ | $h$ | $u$ | $s$ | $c_P$ | $\alpha_P$ | $\lambda$ | $\eta$ |
|---|---|---|---|---|---|---|---|---|
| °C | dm³/kg | kJ/kg | kJ/kg | kJ/kg K | kJ/kg K | $10^{-3}/\mathrm{K}$ | mW/K m | µPa s |
| 0.0 | 0.9992 | 2.0 | 0.0005 | −0.000 | 4.218 | −0.072 | 562.1 | 1787 |
| 5.0 | 0.9991 | 23.0 | 21.0 | 0.076 | 4.191 | 0.017 | 571.5 | 1515 |
| 10.0 | 0.9994 | 43.9 | 41.9 | 0.151 | 4.180 | 0.092 | 581.0 | 1304 |
| 15.0 | 1.0000 | 64.8 | 62.8 | 0.224 | 4.177 | 0.156 | 590.3 | 1136 |
| 20.0 | 1.0009 | 85.7 | 83.7 | 0.296 | 4.177 | 0.212 | 599.3 | 1001 |
| 25.0 | 1.0021 | 106.6 | 104.6 | 0.366 | 4.178 | 0.261 | 608.0 | 890.0 |
| 30.0 | 1.0035 | 127.5 | 125.5 | 0.436 | 4.178 | 0.307 | 616.3 | 797.5 |
| 35.0 | 1.0052 | 148.4 | 146.4 | 0.504 | 4.178 | 0.348 | 624.1 | 719.6 |
| 40.0 | 1.0070 | 169.3 | 167.3 | 0.572 | 4.178 | 0.386 | 631.4 | 653.4 |
| 45.0 | 1.0090 | 190.2 | 188.1 | 0.638 | 4.177 | 0.423 | 638.2 | 596.6 |
| 50.0 | 1.0113 | 211.0 | 209.0 | 0.703 | 4.177 | 0.457 | 644.4 | 547.4 |
| 60.0 | 1.0162 | 252.8 | 250.8 | 0.830 | 4.178 | 0.521 | 655.3 | 467.0 |
| 70.0 | 1.0218 | 294.6 | 292.6 | 0.954 | 4.183 | 0.581 | 664.0 | 404.5 |
| 80.0 | 1.0281 | 336.5 | 334.4 | 1.074 | 4.190 | 0.638 | 671.0 | 355.0 |
| 90.0 | 1.0350 | 378.4 | 376.4 | 1.191 | 4.200 | 0.693 | 676.3 | 315.1 |
| 100.0 | 1.0424 | 420.5 | 418.4 | 1.305 | 4.213 | 0.747 | 680.2 | 282.4 |
| 125.0 | 1.0639 | 526.3 | 524.2 | 1.580 | 4.253 | 0.880 | 684.7 | 222.7 |
| 150.0 | 1.0894 | 633.3 | 631.1 | 1.840 | 4.307 | 1.021 | 683.1 | 182.9 |
| 175.0 | 1.1197 | 741.8 | 739.6 | 2.090 | 4.381 | 1.180 | 676.2 | 155.0 |
| 200.0 | 1.1560 | 852.6 | 850.2 | 2.330 | 4.486 | 1.374 | 663.8 | 134.5 |
| 212.42 | | | | saturation | | | | |
| liquid | 1.1767 | 908.7 | 906.3 | 2.447 | 4.556 | 1.490 | 655.4 | 126.1 |
| vapour | 99.588 | 2798 | 2599 | 6.340 | 2.981 | 3.409 | 42.57 | 16.14 |
| 225.0 | 103.74 | 2834 | 2627 | 6.413 | 2.794 | 3.091 | 42.65 | 16.72 |
| 250.0 | 111.41 | 2901 | 2678 | 6.544 | 2.555 | 2.653 | 43.49 | 17.85 |
| 275.0 | 118.59 | 2963 | 2726 | 6.660 | 2.414 | 2.360 | 44.94 | 18.97 |
| 300.0 | 125.45 | 3022 | 2771 | 6.765 | 2.328 | 2.147 | 46.81 | 20.07 |
| 325.0 | 132.09 | 3080 | 2816 | 6.863 | 2.273 | 1.983 | 48.98 | 21.17 |
| 350.0 | 138.56 | 3136 | 2859 | 6.956 | 2.239 | 1.851 | 51.37 | 22.25 |
| 375.0 | 144.92 | 3192 | 2902 | 7.043 | 2.217 | 1.741 | 53.93 | 23.33 |
| 400.0 | 151.19 | 3247 | 2945 | 7.127 | 2.205 | 1.648 | 56.62 | 24.40 |
| 450.0 | 163.52 | 3357 | 3030 | 7.285 | 2.199 | 1.495 | 62.31 | 26.51 |
| 500.0 | 175.67 | 3467 | 3116 | 7.432 | 2.208 | 1.374 | 68.33 | 28.60 |
| 550.0 | 187.68 | 3578 | 3203 | 7.571 | 2.225 | 1.275 | 74.60 | 30.66 |
| 600.0 | 199.60 | 3690 | 3291 | 7.702 | 2.247 | 1.190 | 81.07 | 32.68 |
| 650.0 | 211.45 | 3803 | 3380 | 7.828 | 2.274 | 1.118 | 87.70 | 34.67 |
| 700.0 | 223.25 | 3917 | 3471 | 7.949 | 2.303 | 1.055 | 94.45 | 36.63 |
| 750.0 | 235.00 | 4033 | 3563 | 8.065 | 2.333 | 0.999 | 101.3 | 38.56 |
| 800.0 | 246.73 | 4150 | 3657 | 8.177 | 2.364 | 0.949 | 108.3 | 40.46 |
| 850.0 | 258.42 | 4269 | 3753 | 8.286 | 2.396 | 0.905 | 115.3 | 42.32 |
| 900.0 | 270.10 | 4390 | 3850 | 8.391 | 2.427 | 0.864 | 122.4 | 44.15 |
| 950.0 | 281.76 | 4512 | 3949 | 8.492 | 2.459 | 0.827 | 129.6 | 45.96 |
| 1000.0 | 293.41 | 4636 | 4049 | 8.592 | 2.490 | 0.793 | 136.8 | 47.73 |

Table 1 (cont.)

$$p = 2.5\,\mathrm{MPa} = 25\ \mathrm{bar}$$

| $t$ | $v$ | $h$ | $u$ | $s$ | $c_\mathrm{p}$ | $\alpha_\mathrm{p}$ | $\lambda$ | $\eta$ |
|---|---|---|---|---|---|---|---|---|
| °C | dm³/kg | kJ/kg | kJ/kg | kJ/kg K | kJ/kg K | $10^{-3}$/K | mW/K m | $\mu$Pa s |
| 0.0 | 0.9990 | 2.5 | 0.01 | 0.00003 | 4.215 | −0.070 | 562.4 | 1786 |
| 5.0 | 0.9988 | 23.5 | 21.0 | 0.076 | 4.189 | 0.019 | 571.8 | 1514 |
| 10.0 | 0.9991 | 44.4 | 41.9 | 0.151 | 4.178 | 0.093 | 581.2 | 1303 |
| 15.0 | 0.9997 | 65.3 | 62.8 | 0.224 | 4.176 | 0.157 | 590.5 | 1136 |
| 20.0 | 1.0007 | 86.2 | 83.7 | 0.296 | 4.176 | 0.212 | 599.5 | 1000 |
| 25.0 | 1.0019 | 107.1 | 104.6 | 0.366 | 4.176 | 0.262 | 608.3 | 889.9 |
| 30.0 | 1.0033 | 127.9 | 125.4 | 0.436 | 4.177 | 0.307 | 616.5 | 797.5 |
| 35.0 | 1.0049 | 148.8 | 146.3 | 0.504 | 4.177 | 0.348 | 624.4 | 719.6 |
| 40.0 | 1.0068 | 169.7 | 167.2 | 0.571 | 4.176 | 0.387 | 631.7 | 653.4 |
| 45.0 | 1.0088 | 190.6 | 188.1 | 0.637 | 4.176 | 0.423 | 638.4 | 596.6 |
| 50.0 | 1.0110 | 211.5 | 208.9 | 0.703 | 4.176 | 0.457 | 644.7 | 547.5 |
| 60.0 | 1.0160 | 253.2 | 250.7 | 0.830 | 4.177 | 0.521 | 655.5 | 467.1 |
| 70.0 | 1.0216 | 295.0 | 292.5 | 0.953 | 4.182 | 0.581 | 664.3 | 404.7 |
| 80.0 | 1.0279 | 336.9 | 334.3 | 1.074 | 4.189 | 0.638 | 671.2 | 355.1 |
| 90.0 | 1.0347 | 378.8 | 376.2 | 1.191 | 4.199 | 0.692 | 676.5 | 315.2 |
| 100.0 | 1.0422 | 420.9 | 418.3 | 1.305 | 4.212 | 0.746 | 680.4 | 282.5 |
| 125.0 | 1.0636 | 526.6 | 524.0 | 1.579 | 4.252 | 0.879 | 685.0 | 222.8 |
| 150.0 | 1.0891 | 633.6 | 630.8 | 1.840 | 4.305 | 1.019 | 683.4 | 183.1 |
| 175.0 | 1.1193 | 742.1 | 739.3 | 2.089 | 4.379 | 1.177 | 676.6 | 155.2 |
| 200.0 | 1.1554 | 852.8 | 849.9 | 2.329 | 4.484 | 1.370 | 664.2 | 134.6 |
| 223.99 | | | | saturation | | | | |
| liquid | 1.1973 | 962.0 | 959.0 | 2.554 | 4.631 | 1.610 | 646.6 | 119.3 |
| vapour | 79.949 | 2802 | 2602 | 6.256 | 3.182 | 3.608 | 45.01 | 16.55 |
| 225.0 | 80.239 | 2805 | 2604 | 6.262 | 3.160 | 3.573 | 45.00 | 16.60 |
| 250.0 | 86.978 | 2879 | 2661 | 6.407 | 2.777 | 2.937 | 45.16 | 17.76 |
| 275.0 | 93.115 | 2945 | 2712 | 6.531 | 2.563 | 2.545 | 46.21 | 18.90 |
| 300.0 | 98.882 | 3007 | 2760 | 6.642 | 2.433 | 2.277 | 47.82 | 20.02 |
| 325.0 | 104.40 | 3067 | 2806 | 6.744 | 2.352 | 2.078 | 49.80 | 21.13 |
| 350.0 | 109.75 | 3125 | 2851 | 6.840 | 2.299 | 1.923 | 52.06 | 22.23 |
| 375.0 | 114.97 | 3182 | 2895 | 6.929 | 2.266 | 1.798 | 54.52 | 23.31 |
| 400.0 | 120.09 | 3239 | 2938 | 7.015 | 2.245 | 1.693 | 57.15 | 24.39 |
| 450.0 | 130.13 | 3350 | 3025 | 7.175 | 2.227 | 1.526 | 62.75 | 26.52 |
| 500.0 | 139.97 | 3462 | 3112 | 7.323 | 2.228 | 1.396 | 68.71 | 28.61 |
| 550.0 | 149.68 | 3573 | 3199 | 7.463 | 2.240 | 1.290 | 74.94 | 30.67 |
| 600.0 | 159.30 | 3686 | 3288 | 7.596 | 2.259 | 1.202 | 81.39 | 32.70 |
| 650.0 | 168.84 | 3799 | 3377 | 7.722 | 2.283 | 1.127 | 88.01 | 34.69 |
| 700.0 | 178.33 | 3914 | 3468 | 7.844 | 2.310 | 1.062 | 94.75 | 36.66 |
| 750.0 | 187.78 | 4030 | 3561 | 7.960 | 2.339 | 1.005 | 101.6 | 38.58 |
| 800.0 | 197.20 | 4148 | 3655 | 8.072 | 2.370 | 0.954 | 108.5 | 40.48 |
| 850.0 | 206.59 | 4267 | 3751 | 8.181 | 2.400 | 0.908 | 115.5 | 42.34 |
| 900.0 | 215.96 | 4388 | 3848 | 8.286 | 2.431 | 0.867 | 122.6 | 44.17 |
| 950.0 | 225.31 | 4510 | 3947 | 8.388 | 2.462 | 0.830 | 129.8 | 45.98 |
| 1000.0 | 234.65 | 4634 | 4048 | 8.488 | 2.493 | 0.795 | 137.0 | 47.75 |

Table 1 (cont.)

$$p = 3.0\,\text{MPa} = 30\ \text{bar}$$

| $t$ | $v$ | $h$ | $u$ | $s$ | $c_\text{p}$ | $\alpha_\text{p}$ | $\lambda$ | $\eta$ |
|---|---|---|---|---|---|---|---|---|
| °C | dm³/kg | kJ/kg | kJ/kg | kJ/kg K | kJ/kg K | $10^{-3}/$K | mW/K m | $\mu$Pa s |
| 0.0 | 0.9987 | 3.0 | 0.02 | 0.00007 | 4.212 | −0.068 | 562.6 | 1785 |
| 5.0 | 0.9986 | 24.0 | 21.0 | 0.076 | 4.186 | 0.021 | 572.1 | 1513 |
| 10.0 | 0.9989 | 44.9 | 41.9 | 0.151 | 4.176 | 0.095 | 581.5 | 1303 |
| 15.0 | 0.9995 | 65.8 | 62.8 | 0.224 | 4.174 | 0.158 | 590.7 | 1135 |
| 20.0 | 1.0004 | 86.7 | 83.6 | 0.296 | 4.174 | 0.213 | 599.8 | 1000 |
| 25.0 | 1.0016 | 107.5 | 104.5 | 0.366 | 4.175 | 0.263 | 608.5 | 889.8 |
| 30.0 | 1.0031 | 128.4 | 125.4 | 0.436 | 4.175 | 0.307 | 616.8 | 797.4 |
| 35.0 | 1.0047 | 149.3 | 146.3 | 0.504 | 4.175 | 0.348 | 624.6 | 719.6 |
| 40.0 | 1.0066 | 170.2 | 167.1 | 0.571 | 4.175 | 0.387 | 631.9 | 653.5 |
| 45.0 | 1.0086 | 191.0 | 188.0 | 0.637 | 4.175 | 0.423 | 638.7 | 596.7 |
| 50.0 | 1.0108 | 211.9 | 208.9 | 0.702 | 4.175 | 0.457 | 644.9 | 547.6 |
| 60.0 | 1.0158 | 253.7 | 250.6 | 0.830 | 4.176 | 0.521 | 655.7 | 467.2 |
| 70.0 | 1.0214 | 295.4 | 292.4 | 0.953 | 4.181 | 0.580 | 664.5 | 404.8 |
| 80.0 | 1.0276 | 337.3 | 334.2 | 1.073 | 4.188 | 0.637 | 671.5 | 355.3 |
| 90.0 | 1.0345 | 379.2 | 376.1 | 1.190 | 4.198 | 0.692 | 676.8 | 315.3 |
| 100.0 | 1.0419 | 421.2 | 418.1 | 1.305 | 4.210 | 0.745 | 680.7 | 282.7 |
| 125.0 | 1.0633 | 527.0 | 523.8 | 1.579 | 4.251 | 0.878 | 685.3 | 222.9 |
| 150.0 | 1.0887 | 633.9 | 630.6 | 1.839 | 4.304 | 1.017 | 683.8 | 183.2 |
| 175.0 | 1.1189 | 742.3 | 739.0 | 2.088 | 4.377 | 1.175 | 676.9 | 155.3 |
| 200.0 | 1.1549 | 853.0 | 849.5 | 2.328 | 4.481 | 1.366 | 664.7 | 134.7 |
| 225.0 | 1.1986 | 966.8 | 963.2 | 2.563 | 4.634 | 1.616 | 646.2 | 118.9 |
| 233.89 | | | | saturation | | | | |
| liquid | 1.2166 | 1008 | 1004 | 2.645 | 4.706 | 1.727 | 637.9 | 114.0 |
| vapour | 66.662 | 2803 | 2603 | 6.186 | 3.381 | 3.815 | 47.25 | 16.90 |
| 250.0 | 70.564 | 2854 | 2643 | 6.286 | 3.047 | 3.284 | 47.00 | 17.67 |
| 275.0 | 76.060 | 2926 | 2698 | 6.420 | 2.733 | 2.759 | 47.57 | 18.83 |
| 300.0 | 81.126 | 2992 | 2749 | 6.537 | 2.550 | 2.421 | 48.87 | 19.97 |
| 325.0 | 85.918 | 3054 | 2797 | 6.644 | 2.437 | 2.182 | 50.66 | 21.09 |
| 350.0 | 90.520 | 3114 | 2843 | 6.742 | 2.364 | 2.001 | 52.78 | 22.20 |
| 375.0 | 94.987 | 3173 | 2888 | 6.834 | 2.317 | 1.858 | 55.14 | 23.30 |
| 400.0 | 99.353 | 3230 | 2932 | 6.921 | 2.286 | 1.740 | 57.69 | 24.38 |
| 450.0 | 107.87 | 3344 | 3020 | 7.083 | 2.255 | 1.557 | 63.19 | 26.52 |
| 500.0 | 116.18 | 3456 | 3108 | 7.234 | 2.248 | 1.418 | 69.10 | 28.62 |
| 550.0 | 124.35 | 3569 | 3196 | 7.375 | 2.256 | 1.306 | 75.30 | 30.69 |
| 600.0 | 132.43 | 3682 | 3285 | 7.508 | 2.272 | 1.214 | 81.73 | 32.72 |
| 650.0 | 140.44 | 3796 | 3375 | 7.635 | 2.293 | 1.136 | 88.33 | 34.72 |
| 700.0 | 148.39 | 3911 | 3466 | 7.757 | 2.318 | 1.069 | 95.06 | 36.68 |
| 750.0 | 156.30 | 4028 | 3559 | 7.874 | 2.346 | 1.010 | 101.9 | 38.61 |
| 800.0 | 164.18 | 4146 | 3653 | 7.987 | 2.375 | 0.958 | 108.8 | 40.50 |
| 850.0 | 172.03 | 4265 | 3749 | 8.095 | 2.405 | 0.912 | 115.8 | 42.37 |
| 900.0 | 179.87 | 4386 | 3847 | 8.201 | 2.435 | 0.870 | 122.8 | 44.20 |
| 950.0 | 187.68 | 4509 | 3946 | 8.303 | 2.466 | 0.832 | 129.9 | 46.00 |
| 1000.0 | 195.48 | 4633 | 4046 | 8.402 | 2.496 | 0.797 | 137.1 | 47.77 |

Table 1 (cont.) $p = 3.5\,\mathrm{MPa} = 35\ \mathrm{bar}$

| $t$ | $v$ | $h$ | $u$ | $s$ | $c_\mathrm{p}$ | $\alpha_\mathrm{p}$ | $\lambda$ | $\eta$ |
|---|---|---|---|---|---|---|---|---|
| °C | dm³/kg | kJ/kg | kJ/kg | kJ/kg K | kJ/kg K | $10^{-3}$/K | mW/K m | $\mu$Pa s |
| 0.0 | 0.9984 | 3.5 | 0.03 | 0.0001 | 4.210 | −0.066 | 562.9 | 1783 |
| 5.0 | 0.9983 | 24.5 | 21.0 | 0.076 | 4.184 | 0.022 | 572.3 | 1513 |
| 10.0 | 0.9986 | 45.4 | 41.9 | 0.151 | 4.175 | 0.096 | 581.7 | 1302 |
| 15.0 | 0.9993 | 66.3 | 62.8 | 0.224 | 4.172 | 0.159 | 591.0 | 1135 |
| 20.0 | 1.0002 | 87.1 | 83.6 | 0.295 | 4.172 | 0.214 | 600.0 | 1000 |
| 25.0 | 1.0014 | 108.0 | 104.5 | 0.366 | 4.173 | 0.263 | 608.7 | 889.7 |
| 30.0 | 1.0028 | 128.9 | 125.3 | 0.435 | 4.174 | 0.308 | 617.0 | 797.4 |
| 35.0 | 1.0045 | 149.7 | 146.2 | 0.504 | 4.174 | 0.349 | 624.8 | 719.6 |
| 40.0 | 1.0063 | 170.6 | 167.1 | 0.571 | 4.174 | 0.387 | 632.1 | 653.5 |
| 45.0 | 1.0084 | 191.5 | 187.9 | 0.637 | 4.174 | 0.423 | 638.9 | 596.8 |
| 50.0 | 1.0106 | 212.3 | 208.8 | 0.702 | 4.174 | 0.457 | 645.1 | 547.7 |
| 60.0 | 1.0156 | 254.1 | 250.5 | 0.829 | 4.175 | 0.520 | 656.0 | 467.4 |
| 70.0 | 1.0212 | 295.8 | 292.3 | 0.953 | 4.179 | 0.580 | 664.8 | 404.9 |
| 80.0 | 1.0274 | 337.7 | 334.1 | 1.073 | 4.187 | 0.636 | 671.7 | 355.4 |
| 90.0 | 1.0342 | 379.6 | 376.0 | 1.190 | 4.197 | 0.691 | 677.1 | 315.5 |
| 100.0 | 1.0417 | 421.6 | 418.0 | 1.304 | 4.209 | 0.744 | 681.0 | 282.8 |
| 125.0 | 1.0630 | 527.3 | 523.6 | 1.578 | 4.249 | 0.876 | 685.6 | 223.0 |
| 150.0 | 1.0884 | 634.2 | 630.4 | 1.839 | 4.302 | 1.015 | 684.1 | 183.3 |
| 175.0 | 1.1185 | 742.6 | 738.7 | 2.088 | 4.375 | 1.172 | 677.3 | 155.4 |
| 200.0 | 1.1544 | 853.2 | 849.1 | 2.328 | 4.478 | 1.363 | 665.1 | 134.9 |
| 225.0 | 1.1980 | 966.9 | 962.7 | 2.562 | 4.630 | 1.611 | 646.7 | 119.0 |
| 242.60 | saturation | | | | | | | |
| liquid | 1.2348 | 1049 | 1045 | 2.725 | 4.783 | 1.844 | 629.4 | 109.7 |
| vapour | 57.054 | 2802 | 2602 | 6.124 | 3.581 | 4.031 | 49.34 | 17.22 |
| 250.0 | 58.712 | 2828 | 2622 | 6.174 | 3.379 | 3.719 | 49.05 | 17.58 |
| 275.0 | 63.809 | 2906 | 2683 | 6.320 | 2.931 | 3.010 | 49.04 | 18.77 |
| 300.0 | 68.403 | 2976 | 2737 | 6.444 | 2.681 | 2.583 | 50.00 | 19.92 |
| 325.0 | 72.688 | 3041 | 2787 | 6.555 | 2.530 | 2.295 | 51.56 | 21.06 |
| 350.0 | 76.768 | 3103 | 2834 | 6.657 | 2.433 | 2.084 | 53.53 | 22.18 |
| 375.0 | 80.702 | 3163 | 2880 | 6.751 | 2.370 | 1.921 | 55.78 | 23.28 |
| 400.0 | 84.529 | 3222 | 2926 | 6.840 | 2.328 | 1.790 | 58.25 | 24.37 |
| 450.0 | 91.956 | 3337 | 3015 | 7.005 | 2.284 | 1.590 | 63.65 | 26.52 |
| 500.0 | 99.175 | 3450 | 3103 | 7.157 | 2.269 | 1.440 | 69.49 | 28.63 |
| 550.0 | 106.25 | 3564 | 3192 | 7.300 | 2.271 | 1.322 | 75.66 | 30.71 |
| 600.0 | 113.24 | 3678 | 3281 | 7.434 | 2.284 | 1.226 | 82.07 | 32.74 |
| 650.0 | 120.15 | 3792 | 3372 | 7.561 | 2.303 | 1.145 | 88.65 | 34.74 |
| 700.0 | 127.00 | 3908 | 3464 | 7.684 | 2.326 | 1.076 | 95.37 | 36.70 |
| 750.0 | 133.82 | 4025 | 3557 | 7.801 | 2.353 | 1.016 | 102.2 | 38.63 |
| 800.0 | 140.60 | 4143 | 3651 | 7.914 | 2.381 | 0.963 | 109.1 | 40.53 |
| 850.0 | 147.35 | 4263 | 3748 | 8.023 | 2.410 | 0.915 | 116.0 | 42.39 |
| 900.0 | 154.09 | 4384 | 3845 | 8.128 | 2.439 | 0.873 | 123.1 | 44.22 |
| 950.0 | 160.80 | 4507 | 3944 | 8.231 | 2.469 | 0.834 | 130.1 | 46.02 |
| 1000.0 | 167.50 | 4631 | 4045 | 8.330 | 2.499 | 0.799 | 137.3 | 47.78 |

Table 1 (cont.)

$$p = 4.0\,\text{MPa} = 40\ \text{bar}$$

| $t$ | $v$ | $h$ | $u$ | $s$ | $c_p$ | $\alpha_p$ | $\lambda$ | $\eta$ |
|---|---|---|---|---|---|---|---|---|
| °C | dm³/kg | kJ/kg | kJ/kg | kJ/kg K | kJ/kg K | $10^{-3}$/K | mW/K m | $\mu$Pa s |
| 0.0 | 0.9982 | 4.0 | 0.04 | 0.0001 | 4.207 | −0.064 | 563.2 | 1782 |
| 5.0 | 0.9981 | 25.0 | 21.0 | 0.076 | 4.182 | 0.024 | 572.6 | 1512 |
| 10.0 | 0.9984 | 45.9 | 41.9 | 0.151 | 4.173 | 0.097 | 581.9 | 1302 |
| 15.0 | 0.9991 | 66.7 | 62.7 | 0.224 | 4.170 | 0.160 | 591.2 | 1135 |
| 20.0 | 1.0000 | 87.6 | 83.6 | 0.295 | 4.171 | 0.215 | 600.2 | 1000 |
| 25.0 | 1.0012 | 108.4 | 104.4 | 0.366 | 4.172 | 0.264 | 608.9 | 889.6 |
| 30.0 | 1.0026 | 129.3 | 125.3 | 0.435 | 4.173 | 0.308 | 617.2 | 797.3 |
| 35.0 | 1.0043 | 150.2 | 146.2 | 0.504 | 4.173 | 0.349 | 625.0 | 719.7 |
| 40.0 | 1.0061 | 171.0 | 167.0 | 0.571 | 4.173 | 0.387 | 632.4 | 653.6 |
| 45.0 | 1.0082 | 191.9 | 187.9 | 0.637 | 4.173 | 0.423 | 639.1 | 596.8 |
| 50.0 | 1.0104 | 212.8 | 208.7 | 0.702 | 4.173 | 0.457 | 645.4 | 547.8 |
| 60.0 | 1.0153 | 254.5 | 250.4 | 0.829 | 4.174 | 0.520 | 656.2 | 467.5 |
| 70.0 | 1.0209 | 296.3 | 292.2 | 0.953 | 4.178 | 0.580 | 665.0 | 405.0 |
| 80.0 | 1.0272 | 338.1 | 334.0 | 1.073 | 4.186 | 0.636 | 672.0 | 355.5 |
| 90.0 | 1.0340 | 380.0 | 375.8 | 1.190 | 4.196 | 0.690 | 677.3 | 315.6 |
| 100.0 | 1.0414 | 422.0 | 417.8 | 1.304 | 4.208 | 0.743 | 681.2 | 282.9 |
| 125.0 | 1.0627 | 527.7 | 523.4 | 1.578 | 4.248 | 0.875 | 685.9 | 223.2 |
| 150.0 | 1.0881 | 634.5 | 630.1 | 1.838 | 4.301 | 1.014 | 684.5 | 183.4 |
| 175.0 | 1.1181 | 742.9 | 738.4 | 2.087 | 4.373 | 1.170 | 677.7 | 155.5 |
| 200.0 | 1.1539 | 853.4 | 848.8 | 2.327 | 4.475 | 1.359 | 665.5 | 135.0 |
| 225.0 | 1.1973 | 967.0 | 962.2 | 2.561 | 4.626 | 1.605 | 647.2 | 119.1 |
| 250.0 | 1.2514 | 1085 | 1080 | 2.793 | 4.856 | 1.954 | 621.4 | 106.2 |
| 250.39 | | | | saturation | | | | |
| liquid | 1.2524 | 1087 | 1082 | 2.796 | 4.861 | 1.961 | 620.9 | 106.0 |
| vapour | 49.771 | 2800 | 2601 | 6.069 | 3.783 | 4.257 | 51.33 | 17.51 |
| 275.0 | 54.553 | 2885 | 2666 | 6.227 | 3.163 | 3.306 | 50.63 | 18.70 |
| 300.0 | 58.822 | 2959 | 2724 | 6.360 | 2.828 | 2.766 | 51.19 | 19.88 |
| 325.0 | 62.742 | 3027 | 2776 | 6.476 | 2.631 | 2.418 | 52.51 | 21.03 |
| 350.0 | 66.437 | 3091 | 2826 | 6.581 | 2.507 | 2.173 | 54.31 | 22.16 |
| 375.0 | 69.977 | 3153 | 2873 | 6.678 | 2.427 | 1.988 | 56.45 | 23.27 |
| 400.0 | 73.403 | 3213 | 2919 | 6.769 | 2.373 | 1.842 | 58.83 | 24.37 |
| 450.0 | 80.019 | 3330 | 3010 | 6.936 | 2.314 | 1.623 | 64.12 | 26.53 |
| 500.0 | 86.421 | 3445 | 3099 | 7.090 | 2.291 | 1.463 | 69.90 | 28.65 |
| 550.0 | 92.681 | 3559 | 3189 | 7.234 | 2.288 | 1.339 | 76.03 | 30.72 |
| 600.0 | 98.841 | 3674 | 3278 | 7.369 | 2.297 | 1.238 | 82.41 | 32.76 |
| 650.0 | 104.93 | 3789 | 3369 | 7.497 | 2.313 | 1.155 | 88.98 | 34.76 |
| 700.0 | 110.96 | 3905 | 3461 | 7.620 | 2.335 | 1.083 | 95.69 | 36.73 |
| 750.0 | 116.95 | 4023 | 3555 | 7.737 | 2.359 | 1.021 | 102.5 | 38.66 |
| 800.0 | 122.91 | 4141 | 3650 | 7.850 | 2.386 | 0.967 | 109.4 | 40.55 |
| 850.0 | 128.84 | 4261 | 3746 | 7.960 | 2.415 | 0.919 | 116.3 | 42.41 |
| 900.0 | 134.75 | 4383 | 3844 | 8.065 | 2.444 | 0.876 | 123.3 | 44.24 |
| 950.0 | 140.64 | 4506 | 3943 | 8.168 | 2.473 | 0.837 | 130.3 | 46.04 |
| 1000.0 | 146.52 | 4630 | 4044 | 8.268 | 2.502 | 0.801 | 137.4 | 47.80 |

Table 1 (cont.)

$$p = 4.5\,\text{MPa} = 45\;\text{bar}$$

| $t$ | $v$ | $h$ | $u$ | $s$ | $c_\text{p}$ | $\alpha_\text{p}$ | $\lambda$ | $\eta$ |
|---|---|---|---|---|---|---|---|---|
| °C | dm³/kg | kJ/kg | kJ/kg | kJ/kg K | kJ/kg K | $10^{-3}$/K | mW/K m | $\mu$Pa s |
| 0.0 | 0.9979 | 4.5 | 0.05 | 0.0002 | 4.204 | −0.062 | 563.5 | 1781 |
| 5.0 | 0.9979 | 25.5 | 21.0 | 0.076 | 4.180 | 0.025 | 572.8 | 1511 |
| 10.0 | 0.9982 | 46.4 | 41.9 | 0.151 | 4.171 | 0.098 | 582.2 | 1301 |
| 15.0 | 0.9988 | 67.2 | 62.7 | 0.224 | 4.169 | 0.161 | 591.4 | 1134 |
| 20.0 | 0.9998 | 88.1 | 83.6 | 0.295 | 4.169 | 0.215 | 600.5 | 1000 |
| 25.0 | 1.0010 | 108.9 | 104.4 | 0.366 | 4.171 | 0.264 | 609.2 | 889.5 |
| 30.0 | 1.0024 | 129.8 | 125.3 | 0.435 | 4.171 | 0.308 | 617.5 | 797.3 |
| 35.0 | 1.0041 | 150.6 | 146.1 | 0.503 | 4.172 | 0.349 | 625.3 | 719.7 |
| 40.0 | 1.0059 | 171.5 | 167.0 | 0.571 | 4.172 | 0.387 | 632.6 | 653.6 |
| 45.0 | 1.0079 | 192.3 | 187.8 | 0.637 | 4.172 | 0.423 | 639.4 | 596.9 |
| 50.0 | 1.0102 | 213.2 | 208.6 | 0.702 | 4.172 | 0.457 | 645.6 | 547.9 |
| 60.0 | 1.0151 | 254.9 | 250.3 | 0.829 | 4.173 | 0.520 | 656.5 | 467.6 |
| 70.0 | 1.0207 | 296.7 | 292.1 | 0.952 | 4.177 | 0.579 | 665.3 | 405.2 |
| 80.0 | 1.0269 | 338.5 | 333.9 | 1.072 | 4.185 | 0.635 | 672.3 | 355.7 |
| 90.0 | 1.0338 | 380.4 | 375.7 | 1.189 | 4.195 | 0.690 | 677.6 | 315.7 |
| 100.0 | 1.0412 | 422.4 | 417.7 | 1.303 | 4.207 | 0.743 | 681.5 | 283.1 |
| 125.0 | 1.0624 | 528.0 | 523.2 | 1.577 | 4.247 | 0.874 | 686.2 | 223.3 |
| 150.0 | 1.0877 | 634.8 | 629.9 | 1.838 | 4.299 | 1.012 | 684.8 | 183.6 |
| 175.0 | 1.1177 | 743.1 | 738.1 | 2.086 | 4.370 | 1.167 | 678.1 | 155.7 |
| 200.0 | 1.1534 | 853.6 | 848.4 | 2.326 | 4.472 | 1.355 | 666.0 | 135.1 |
| 225.0 | 1.1966 | 967.1 | 961.7 | 2.560 | 4.621 | 1.599 | 647.8 | 119.2 |
| 250.0 | 1.2505 | 1085 | 1079 | 2.791 | 4.850 | 1.945 | 622.0 | 106.3 |
| 257.47 | | | | saturation | | | | |
| liquid | 1.2694 | 1121 | 1116 | 2.861 | 4.942 | 2.081 | 612.6 | 102.8 |
| vapour | 44.053 | 2797 | 2599 | 6.019 | 3.989 | 4.493 | 53.26 | 17.78 |
| 275.0 | 47.285 | 2862 | 2649 | 6.139 | 3.436 | 3.661 | 52.38 | 18.64 |
| 300.0 | 51.332 | 2942 | 2711 | 6.281 | 2.994 | 2.973 | 52.48 | 19.84 |
| 325.0 | 54.984 | 3013 | 2766 | 6.403 | 2.742 | 2.554 | 53.51 | 21.00 |
| 350.0 | 58.388 | 3079 | 2817 | 6.512 | 2.587 | 2.268 | 55.13 | 22.14 |
| 375.0 | 61.625 | 3143 | 2865 | 6.612 | 2.486 | 2.058 | 57.14 | 23.26 |
| 400.0 | 64.743 | 3204 | 2913 | 6.704 | 2.419 | 1.896 | 59.44 | 24.37 |
| 450.0 | 70.732 | 3323 | 3005 | 6.875 | 2.344 | 1.657 | 64.60 | 26.54 |
| 500.0 | 76.500 | 3439 | 3095 | 7.030 | 2.312 | 1.487 | 70.31 | 28.66 |
| 550.0 | 82.123 | 3554 | 3185 | 7.175 | 2.304 | 1.356 | 76.40 | 30.74 |
| 600.0 | 87.644 | 3670 | 3275 | 7.311 | 2.309 | 1.251 | 82.77 | 32.78 |
| 650.0 | 93.093 | 3786 | 3367 | 7.440 | 2.323 | 1.164 | 89.32 | 34.79 |
| 700.0 | 98.486 | 3902 | 3459 | 7.563 | 2.343 | 1.091 | 96.01 | 36.75 |
| 750.0 | 103.84 | 4020 | 3553 | 7.681 | 2.366 | 1.027 | 102.8 | 38.68 |
| 800.0 | 109.15 | 4139 | 3648 | 7.794 | 2.392 | 0.972 | 109.6 | 40.58 |
| 850.0 | 114.44 | 4259 | 3744 | 7.904 | 2.419 | 0.923 | 116.6 | 42.44 |
| 900.0 | 119.71 | 4381 | 3842 | 8.010 | 2.448 | 0.879 | 123.5 | 44.26 |
| 950.0 | 124.96 | 4504 | 3942 | 8.112 | 2.476 | 0.839 | 130.5 | 46.06 |
| 1000.0 | 130.20 | 4628 | 4043 | 8.212 | 2.505 | 0.803 | 137.6 | 47.82 |

Table 1 (cont.)

$$p = 5.0 \, \text{MPa} = 50 \, \text{bar}$$

| $t$ | $v$ | $h$ | $u$ | $s$ | $c_p$ | $\alpha_p$ | $\lambda$ | $\eta$ |
|---|---|---|---|---|---|---|---|---|
| °C | dm³/kg | kJ/kg | kJ/kg | kJ/kg K | kJ/kg K | $10^{-3}$/K | mW/K m | $\mu$Pa s |
| 0.0 | 0.9977 | 5.0 | 0.06 | 0.0002 | 4.202 | −0.060 | 563.7 | 1780 |
| 5.0 | 0.9976 | 26.0 | 21.0 | 0.076 | 4.178 | 0.027 | 573.1 | 1510 |
| 10.0 | 0.9979 | 46.9 | 41.9 | 0.151 | 4.169 | 0.099 | 582.4 | 1301 |
| 15.0 | 0.9986 | 67.7 | 62.7 | 0.223 | 4.167 | 0.162 | 591.7 | 1134 |
| 20.0 | 0.9995 | 88.5 | 83.5 | 0.295 | 4.168 | 0.216 | 600.7 | 1000.0 |
| 25.0 | 1.0007 | 109.4 | 104.4 | 0.366 | 4.169 | 0.265 | 609.4 | 889.4 |
| 30.0 | 1.0022 | 130.2 | 125.2 | 0.435 | 4.170 | 0.309 | 617.7 | 797.3 |
| 35.0 | 1.0038 | 151.1 | 146.0 | 0.503 | 4.171 | 0.349 | 625.5 | 719.7 |
| 40.0 | 1.0057 | 171.9 | 166.9 | 0.570 | 4.171 | 0.387 | 632.8 | 653.7 |
| 45.0 | 1.0077 | 192.8 | 187.7 | 0.636 | 4.170 | 0.423 | 639.6 | 597.0 |
| 50.0 | 1.0099 | 213.6 | 208.6 | 0.701 | 4.170 | 0.456 | 645.8 | 547.9 |
| 60.0 | 1.0149 | 255.3 | 250.3 | 0.829 | 4.172 | 0.520 | 656.7 | 467.7 |
| 70.0 | 1.0205 | 297.1 | 292.0 | 0.952 | 4.176 | 0.579 | 665.5 | 405.3 |
| 80.0 | 1.0267 | 338.9 | 333.7 | 1.072 | 4.183 | 0.635 | 672.5 | 355.8 |
| 90.0 | 1.0335 | 380.8 | 375.6 | 1.189 | 4.193 | 0.689 | 677.9 | 315.9 |
| 100.0 | 1.0409 | 422.7 | 417.5 | 1.303 | 4.206 | 0.742 | 681.8 | 283.2 |
| 125.0 | 1.0621 | 528.4 | 523.1 | 1.577 | 4.245 | 0.873 | 686.5 | 223.4 |
| 150.0 | 1.0874 | 635.1 | 629.7 | 1.837 | 4.298 | 1.010 | 685.1 | 183.7 |
| 175.0 | 1.1173 | 743.4 | 737.8 | 2.086 | 4.368 | 1.165 | 678.5 | 155.8 |
| 200.0 | 1.1529 | 853.8 | 848.0 | 2.325 | 4.469 | 1.352 | 666.4 | 135.2 |
| 225.0 | 1.1960 | 967.3 | 961.3 | 2.559 | 4.617 | 1.594 | 648.3 | 119.4 |
| 250.0 | 1.2496 | 1085 | 1079 | 2.790 | 4.843 | 1.935 | 622.7 | 106.5 |
| 263.98 | | | | saturation | | | | |
| liquid | 1.2861 | 1154 | 1147 | 2.920 | 5.025 | 2.204 | 604.3 | 100.1 |
| vapour | 39.440 | 2793 | 2596 | 5.973 | 4.200 | 4.741 | 55.16 | 18.03 |
| 275.0 | 41.399 | 2837 | 2630 | 6.053 | 3.766 | 4.094 | 54.32 | 18.58 |
| 300.0 | 45.303 | 2923 | 2696 | 6.207 | 3.181 | 3.210 | 53.86 | 19.80 |
| 325.0 | 48.755 | 2998 | 2754 | 6.335 | 2.863 | 2.704 | 54.57 | 20.97 |
| 350.0 | 51.935 | 3067 | 2808 | 6.448 | 2.672 | 2.371 | 55.99 | 22.12 |
| 375.0 | 54.935 | 3132 | 2858 | 6.551 | 2.549 | 2.133 | 57.87 | 23.25 |
| 400.0 | 57.808 | 3195 | 2906 | 6.646 | 2.468 | 1.953 | 60.06 | 24.37 |
| 450.0 | 63.298 | 3316 | 2999 | 6.819 | 2.376 | 1.693 | 65.10 | 26.55 |
| 500.0 | 68.560 | 3433 | 3091 | 6.976 | 2.335 | 1.511 | 70.74 | 28.68 |
| 550.0 | 73.675 | 3550 | 3181 | 7.122 | 2.320 | 1.373 | 76.79 | 30.76 |
| 600.0 | 78.686 | 3666 | 3272 | 7.259 | 2.322 | 1.263 | 83.13 | 32.81 |
| 650.0 | 83.623 | 3782 | 3364 | 7.388 | 2.333 | 1.174 | 89.67 | 34.81 |
| 700.0 | 88.505 | 3899 | 3457 | 7.512 | 2.351 | 1.098 | 96.34 | 36.78 |
| 750.0 | 93.343 | 4017 | 3551 | 7.630 | 2.373 | 1.033 | 103.1 | 38.71 |
| 800.0 | 98.148 | 4137 | 3646 | 7.744 | 2.398 | 0.976 | 109.9 | 40.60 |
| 850.0 | 102.93 | 4257 | 3742 | 7.854 | 2.424 | 0.926 | 116.8 | 42.46 |
| 900.0 | 107.68 | 4379 | 3841 | 7.960 | 2.452 | 0.882 | 123.8 | 44.29 |
| 950.0 | 112.42 | 4502 | 3940 | 8.063 | 2.480 | 0.842 | 130.7 | 46.08 |
| 1000.0 | 117.15 | 4627 | 4041 | 8.163 | 2.508 | 0.805 | 137.7 | 47.84 |

Table 1 (cont.)

$$p = 6.0\,\text{MPa} = 60\ \text{bar}$$

| $t$ | $v$ | $h$ | $u$ | $s$ | $c_p$ | $\alpha_p$ | $\lambda$ | $\eta$ |
|---|---|---|---|---|---|---|---|---|
| °C | dm³/kg | kJ/kg | kJ/kg | kJ/kg K | kJ/kg K | $10^{-3}$/K | mW/K m | $\mu$Pa s |
| 0.0 | 0.9972 | 6.1 | 0.08 | 0.0003 | 4.197 | −0.056 | 564.3 | 1777 |
| 5.0 | 0.9971 | 27.0 | 21.0 | 0.076 | 4.173 | 0.030 | 573.6 | 1509 |
| 10.0 | 0.9975 | 47.8 | 41.8 | 0.150 | 4.165 | 0.102 | 582.9 | 1300 |
| 15.0 | 0.9981 | 68.6 | 62.7 | 0.223 | 4.164 | 0.163 | 592.1 | 1133 |
| 20.0 | 0.9991 | 89.5 | 83.5 | 0.295 | 4.165 | 0.217 | 601.2 | 999.6 |
| 25.0 | 1.0003 | 110.3 | 104.3 | 0.365 | 4.166 | 0.266 | 609.8 | 889.1 |
| 30.0 | 1.0017 | 131.1 | 125.1 | 0.435 | 4.168 | 0.310 | 618.1 | 797.2 |
| 35.0 | 1.0034 | 152.0 | 145.9 | 0.503 | 4.168 | 0.350 | 626.0 | 719.7 |
| 40.0 | 1.0052 | 172.8 | 166.8 | 0.570 | 4.168 | 0.387 | 633.3 | 653.8 |
| 45.0 | 1.0073 | 193.6 | 187.6 | 0.636 | 4.168 | 0.423 | 640.1 | 597.1 |
| 50.0 | 1.0095 | 214.5 | 208.4 | 0.701 | 4.168 | 0.456 | 646.3 | 548.1 |
| 60.0 | 1.0144 | 256.2 | 250.1 | 0.828 | 4.170 | 0.519 | 657.2 | 467.9 |
| 70.0 | 1.0200 | 297.9 | 291.8 | 0.951 | 4.174 | 0.578 | 666.0 | 405.6 |
| 80.0 | 1.0262 | 339.7 | 333.5 | 1.071 | 4.181 | 0.634 | 673.0 | 356.1 |
| 90.0 | 1.0330 | 381.5 | 375.3 | 1.188 | 4.191 | 0.688 | 678.4 | 316.1 |
| 100.0 | 1.0404 | 423.5 | 417.3 | 1.302 | 4.204 | 0.740 | 682.3 | 283.5 |
| 125.0 | 1.0616 | 529.1 | 522.7 | 1.576 | 4.243 | 0.870 | 687.1 | 223.7 |
| 150.0 | 1.0868 | 635.7 | 629.2 | 1.836 | 4.294 | 1.007 | 685.8 | 183.9 |
| 175.0 | 1.1165 | 743.9 | 737.2 | 2.084 | 4.364 | 1.160 | 679.2 | 156.0 |
| 200.0 | 1.1519 | 854.2 | 847.3 | 2.324 | 4.464 | 1.344 | 667.3 | 135.5 |
| 225.0 | 1.1947 | 967.5 | 960.3 | 2.557 | 4.609 | 1.583 | 649.3 | 119.6 |
| 250.0 | 1.2478 | 1085 | 1077 | 2.788 | 4.830 | 1.917 | 624.0 | 106.7 |
| 275.0 | 1.3170 | 1210 | 1202 | 3.021 | 5.190 | 2.447 | 589.0 | 95.56 |
| 275.62 | | | | saturation | | | | |
| liquid | 1.3190 | 1213 | 1205 | 3.027 | 5.201 | 2.464 | 588.0 | 95.30 |
| vapour | 32.442 | 2783 | 2589 | 5.889 | 4.643 | 5.278 | 58.94 | 18.51 |
| 300.0 | 36.151 | 2883 | 2666 | 6.066 | 3.642 | 3.805 | 57.00 | 19.73 |
| 325.0 | 39.352 | 2967 | 2731 | 6.210 | 3.144 | 3.054 | 56.91 | 20.93 |
| 350.0 | 42.218 | 3042 | 2788 | 6.332 | 2.861 | 2.603 | 57.85 | 22.10 |
| 375.0 | 44.875 | 3111 | 2842 | 6.441 | 2.686 | 2.298 | 59.41 | 23.25 |
| 400.0 | 47.390 | 3177 | 2892 | 6.540 | 2.572 | 2.075 | 61.38 | 24.37 |
| 450.0 | 52.139 | 3301 | 2989 | 6.720 | 2.442 | 1.768 | 66.14 | 26.57 |
| 500.0 | 56.647 | 3422 | 3082 | 6.881 | 2.381 | 1.561 | 71.62 | 28.71 |
| 550.0 | 61.000 | 3540 | 3174 | 7.029 | 2.354 | 1.408 | 77.58 | 30.81 |
| 600.0 | 65.248 | 3658 | 3266 | 7.167 | 2.348 | 1.289 | 83.87 | 32.86 |
| 650.0 | 69.419 | 3775 | 3359 | 7.298 | 2.354 | 1.193 | 90.37 | 34.86 |
| 700.0 | 73.533 | 3893 | 3452 | 7.423 | 2.368 | 1.113 | 97.02 | 36.83 |
| 750.0 | 77.604 | 4012 | 3546 | 7.542 | 2.387 | 1.044 | 103.7 | 38.76 |
| 800.0 | 81.641 | 4132 | 3642 | 7.656 | 2.409 | 0.985 | 110.5 | 40.65 |
| 850.0 | 85.650 | 4253 | 3739 | 7.766 | 2.434 | 0.934 | 117.4 | 42.51 |
| 900.0 | 89.638 | 4375 | 3837 | 7.873 | 2.460 | 0.888 | 124.3 | 44.33 |
| 950.0 | 93.608 | 4499 | 3937 | 7.976 | 2.487 | 0.847 | 131.2 | 46.13 |
| 1000.0 | 97.563 | 4624 | 4039 | 8.076 | 2.514 | 0.809 | 138.1 | 47.89 |

Table 1 (cont.) $p = 7.0\,\text{MPa} = 70\;\text{bar}$

| $t$ | $v$ | $h$ | $u$ | $s$ | $c_p$ | $\alpha_p$ | $\lambda$ | $\eta$ |
|---|---|---|---|---|---|---|---|---|
| °C | dm³/kg | kJ/kg | kJ/kg | kJ/kg K | kJ/kg K | $10^{-3}$/K | mW/K m | μPa s |
| 0.0 | 0.9967 | 7.1 | 0.1 | 0.0003 | 4.192 | −0.052 | 564.8 | 1775 |
| 5.0 | 0.9967 | 28.0 | 21.0 | 0.076 | 4.169 | 0.033 | 574.1 | 1507 |
| 10.0 | 0.9970 | 48.8 | 41.8 | 0.150 | 4.161 | 0.104 | 583.4 | 1299 |
| 15.0 | 0.9977 | 69.6 | 62.6 | 0.223 | 4.160 | 0.165 | 592.6 | 1133 |
| 20.0 | 0.9986 | 90.4 | 83.4 | 0.295 | 4.162 | 0.219 | 601.6 | 999.2 |
| 25.0 | 0.9999 | 111.2 | 104.2 | 0.365 | 4.164 | 0.267 | 610.3 | 888.9 |
| 30.0 | 1.0013 | 132.0 | 125.0 | 0.434 | 4.165 | 0.310 | 618.6 | 797.1 |
| 35.0 | 1.0030 | 152.9 | 145.8 | 0.503 | 4.166 | 0.350 | 626.4 | 719.7 |
| 40.0 | 1.0048 | 173.7 | 166.7 | 0.570 | 4.166 | 0.388 | 633.7 | 653.8 |
| 45.0 | 1.0068 | 194.5 | 187.5 | 0.636 | 4.166 | 0.423 | 640.5 | 597.3 |
| 50.0 | 1.0091 | 215.3 | 208.3 | 0.701 | 4.166 | 0.456 | 646.8 | 548.3 |
| 60.0 | 1.0140 | 257.0 | 249.9 | 0.827 | 4.168 | 0.519 | 657.6 | 468.2 |
| 70.0 | 1.0196 | 298.7 | 291.6 | 0.951 | 4.172 | 0.577 | 666.5 | 405.8 |
| 80.0 | 1.0258 | 340.5 | 333.3 | 1.071 | 4.179 | 0.633 | 673.5 | 356.3 |
| 90.0 | 1.0325 | 382.3 | 375.1 | 1.188 | 4.189 | 0.686 | 678.9 | 316.4 |
| 100.0 | 1.0399 | 424.3 | 417.0 | 1.302 | 4.201 | 0.738 | 682.9 | 283.7 |
| 125.0 | 1.0610 | 529.8 | 522.3 | 1.575 | 4.240 | 0.868 | 687.7 | 224.0 |
| 150.0 | 1.0861 | 636.4 | 628.8 | 1.835 | 4.291 | 1.003 | 686.5 | 184.2 |
| 175.0 | 1.1157 | 744.5 | 736.7 | 2.083 | 4.360 | 1.155 | 680.0 | 156.3 |
| 200.0 | 1.1510 | 854.6 | 846.6 | 2.322 | 4.458 | 1.337 | 668.1 | 135.7 |
| 225.0 | 1.1934 | 967.8 | 959.4 | 2.555 | 4.601 | 1.572 | 650.3 | 119.9 |
| 250.0 | 1.2460 | 1085 | 1076 | 2.785 | 4.817 | 1.899 | 625.2 | 107.0 |
| 275.0 | 1.3143 | 1209 | 1200 | 3.018 | 5.166 | 2.412 | 590.7 | 95.89 |
| 285.86 | | | | saturation | | | | |
| liquid | 1.3515 | 1266 | 1257 | 3.121 | 5.394 | 2.749 | 572.1 | 91.30 |
| vapour | 27.372 | 2771 | 2580 | 5.813 | 5.121 | 5.881 | 62.82 | 18.96 |
| 300.0 | 29.460 | 2837 | 2631 | 5.929 | 4.272 | 4.639 | 60.83 | 19.68 |
| 325.0 | 32.556 | 2933 | 2705 | 6.093 | 3.490 | 3.494 | 59.61 | 20.90 |
| 350.0 | 35.231 | 3015 | 2768 | 6.227 | 3.082 | 2.875 | 59.92 | 22.09 |
| 375.0 | 37.660 | 3088 | 2825 | 6.343 | 2.840 | 2.484 | 61.09 | 23.25 |
| 400.0 | 39.928 | 3157 | 2878 | 6.447 | 2.686 | 2.210 | 62.81 | 24.39 |
| 450.0 | 44.159 | 3287 | 2978 | 6.633 | 2.512 | 1.848 | 67.24 | 26.60 |
| 500.0 | 48.132 | 3410 | 3073 | 6.798 | 2.428 | 1.613 | 72.55 | 28.75 |
| 550.0 | 51.944 | 3530 | 3167 | 6.949 | 2.389 | 1.444 | 78.41 | 30.86 |
| 600.0 | 55.647 | 3649 | 3260 | 7.089 | 2.375 | 1.315 | 84.64 | 32.91 |
| 650.0 | 59.272 | 3768 | 3353 | 7.221 | 2.375 | 1.212 | 91.11 | 34.92 |
| 700.0 | 62.839 | 3887 | 3447 | 7.347 | 2.385 | 1.128 | 97.72 | 36.89 |
| 750.0 | 66.362 | 4007 | 3542 | 7.466 | 2.401 | 1.056 | 104.4 | 38.82 |
| 800.0 | 69.850 | 4127 | 3638 | 7.581 | 2.421 | 0.995 | 111.2 | 40.71 |
| 850.0 | 73.311 | 4249 | 3736 | 7.692 | 2.444 | 0.941 | 117.9 | 42.56 |
| 900.0 | 76.750 | 4372 | 3834 | 7.799 | 2.469 | 0.894 | 124.8 | 44.38 |
| 950.0 | 80.171 | 4496 | 3934 | 7.903 | 2.494 | 0.851 | 131.6 | 46.17 |
| 1000.0 | 83.576 | 4621 | 4036 | 8.003 | 2.521 | 0.813 | 138.5 | 47.93 |

Table 1 (cont.)

$$p = 8.0\,\text{MPa} = 80\ \text{bar}$$

| $t$ | $v$ | $h$ | $u$ | $s$ | $c_p$ | $\alpha_p$ | $\lambda$ | $\eta$ |
|---|---|---|---|---|---|---|---|---|
| °C | dm³/kg | kJ/kg | kJ/kg | kJ/kg K | kJ/kg K | $10^{-3}$/K | mW/K m | µPa s |
| 0.0 | 0.9962 | 8.1 | 0.1 | 0.0004 | 4.186 | −0.048 | 565.4 | 1773 |
| 5.0 | 0.9962 | 29.0 | 21.0 | 0.076 | 4.165 | 0.036 | 574.6 | 1506 |
| 10.0 | 0.9965 | 49.8 | 41.8 | 0.150 | 4.158 | 0.106 | 583.9 | 1298 |
| 15.0 | 0.9972 | 70.5 | 62.6 | 0.223 | 4.157 | 0.167 | 593.1 | 1132 |
| 20.0 | 0.9982 | 91.3 | 83.3 | 0.294 | 4.159 | 0.220 | 602.1 | 998.8 |
| 25.0 | 0.9994 | 112.1 | 104.1 | 0.365 | 4.161 | 0.268 | 610.7 | 888.7 |
| 30.0 | 1.0009 | 132.9 | 124.9 | 0.434 | 4.162 | 0.311 | 619.0 | 797.0 |
| 35.0 | 1.0025 | 153.8 | 145.7 | 0.502 | 4.163 | 0.351 | 626.9 | 719.7 |
| 40.0 | 1.0044 | 174.6 | 166.5 | 0.569 | 4.164 | 0.388 | 634.2 | 653.9 |
| 45.0 | 1.0064 | 195.4 | 187.3 | 0.635 | 4.164 | 0.423 | 641.0 | 597.4 |
| 50.0 | 1.0086 | 216.2 | 208.1 | 0.700 | 4.164 | 0.456 | 647.2 | 548.5 |
| 60.0 | 1.0136 | 257.9 | 249.7 | 0.827 | 4.166 | 0.518 | 658.1 | 468.4 |
| 70.0 | 1.0191 | 299.5 | 291.4 | 0.950 | 4.170 | 0.576 | 667.0 | 406.1 |
| 80.0 | 1.0253 | 341.3 | 333.1 | 1.070 | 4.177 | 0.632 | 674.0 | 356.6 |
| 90.0 | 1.0321 | 383.1 | 374.8 | 1.187 | 4.187 | 0.685 | 679.4 | 316.7 |
| 100.0 | 1.0394 | 425.0 | 416.7 | 1.301 | 4.199 | 0.737 | 683.4 | 284.0 |
| 125.0 | 1.0605 | 530.4 | 522.0 | 1.574 | 4.238 | 0.865 | 688.3 | 224.2 |
| 150.0 | 1.0855 | 637.0 | 628.3 | 1.834 | 4.288 | 1.000 | 687.1 | 184.4 |
| 175.0 | 1.1150 | 745.0 | 736.1 | 2.082 | 4.357 | 1.150 | 680.7 | 156.5 |
| 200.0 | 1.1500 | 855.1 | 845.9 | 2.321 | 4.453 | 1.330 | 669.0 | 136.0 |
| 225.0 | 1.1922 | 968.0 | 958.5 | 2.553 | 4.593 | 1.561 | 651.3 | 120.1 |
| 250.0 | 1.2443 | 1085 | 1075 | 2.783 | 4.804 | 1.882 | 626.5 | 107.3 |
| 275.0 | 1.3116 | 1209 | 1198 | 3.014 | 5.143 | 2.379 | 592.4 | 96.21 |
| 295.04 | saturation | | | | | | | |
| liquid | 1.3843 | 1316 | 1305 | 3.207 | 5.609 | 3.068 | 556.5 | 87.81 |
| vapour | 23.520 | 2757 | 2569 | 5.743 | 5.647 | 6.565 | 66.93 | 19.40 |
| 300.0 | 24.256 | 2784 | 2590 | 5.790 | 5.196 | 5.904 | 65.77 | 19.65 |
| 325.0 | 27.377 | 2896 | 2677 | 5.981 | 3.927 | 4.060 | 62.79 | 20.89 |
| 350.0 | 29.946 | 2986 | 2746 | 6.129 | 3.342 | 3.201 | 62.27 | 22.09 |
| 375.0 | 32.221 | 3065 | 2807 | 6.253 | 3.014 | 2.696 | 62.96 | 23.26 |
| 400.0 | 34.314 | 3138 | 2863 | 6.363 | 2.811 | 2.360 | 64.35 | 24.41 |
| 450.0 | 38.165 | 3272 | 2966 | 6.555 | 2.586 | 1.933 | 68.42 | 26.64 |
| 500.0 | 41.741 | 3398 | 3064 | 6.724 | 2.478 | 1.668 | 73.52 | 28.80 |
| 550.0 | 45.150 | 3520 | 3159 | 6.878 | 2.425 | 1.482 | 79.27 | 30.91 |
| 600.0 | 48.446 | 3641 | 3253 | 7.020 | 2.402 | 1.342 | 85.44 | 32.96 |
| 650.0 | 51.662 | 3761 | 3348 | 7.154 | 2.396 | 1.233 | 91.87 | 34.98 |
| 700.0 | 54.819 | 3881 | 3442 | 7.280 | 2.402 | 1.143 | 98.44 | 36.94 |
| 750.0 | 57.931 | 4001 | 3538 | 7.401 | 2.415 | 1.068 | 105.1 | 38.87 |
| 800.0 | 61.008 | 4122 | 3634 | 7.516 | 2.433 | 1.004 | 111.8 | 40.76 |
| 850.0 | 64.057 | 4245 | 3732 | 7.628 | 2.454 | 0.948 | 118.5 | 42.62 |
| 900.0 | 67.085 | 4368 | 3831 | 7.735 | 2.477 | 0.900 | 125.3 | 44.44 |
| 950.0 | 70.094 | 4492 | 3932 | 7.839 | 2.502 | 0.856 | 132.1 | 46.22 |
| 1000.0 | 73.087 | 4618 | 4033 | 7.940 | 2.527 | 0.817 | 138.9 | 47.97 |

Table 1 (cont.) $p = 9.0\,\text{MPa} = 90\ \text{bar}$

| $t$ | $v$ | $h$ | $u$ | $s$ | $c_p$ | $\alpha_p$ | $\lambda$ | $\eta$ |
|---|---|---|---|---|---|---|---|---|
| °C | dm³/kg | kJ/kg | kJ/kg | kJ/kg K | kJ/kg K | $10^{-3}$/K | mW/K m | $\mu$Pa s |
| 0.0 | 0.9957 | 9.1 | 0.1 | 0.0004 | 4.181 | −0.044 | 566.0 | 1770 |
| 5.0 | 0.9957 | 29.9 | 21.0 | 0.076 | 4.160 | 0.039 | 575.1 | 1504 |
| 10.0 | 0.9961 | 50.7 | 41.8 | 0.150 | 4.154 | 0.109 | 584.4 | 1297 |
| 15.0 | 0.9968 | 71.5 | 62.5 | 0.223 | 4.154 | 0.169 | 593.5 | 1132 |
| 20.0 | 0.9977 | 92.3 | 83.3 | 0.294 | 4.156 | 0.222 | 602.5 | 998.5 |
| 25.0 | 0.9990 | 113.0 | 104.1 | 0.365 | 4.158 | 0.269 | 611.2 | 888.6 |
| 30.0 | 1.0004 | 133.8 | 124.8 | 0.434 | 4.160 | 0.312 | 619.5 | 797.0 |
| 35.0 | 1.0021 | 154.6 | 145.6 | 0.502 | 4.161 | 0.351 | 627.3 | 719.8 |
| 40.0 | 1.0039 | 175.5 | 166.4 | 0.569 | 4.161 | 0.388 | 634.6 | 654.0 |
| 45.0 | 1.0060 | 196.3 | 187.2 | 0.635 | 4.161 | 0.423 | 641.4 | 597.6 |
| 50.0 | 1.0082 | 217.1 | 208.0 | 0.700 | 4.162 | 0.456 | 647.7 | 548.7 |
| 60.0 | 1.0131 | 258.7 | 249.6 | 0.826 | 4.163 | 0.518 | 658.6 | 468.6 |
| 70.0 | 1.0187 | 300.3 | 291.2 | 0.950 | 4.168 | 0.576 | 667.5 | 406.3 |
| 80.0 | 1.0248 | 342.1 | 332.8 | 1.069 | 4.175 | 0.631 | 674.5 | 356.9 |
| 90.0 | 1.0316 | 383.9 | 374.6 | 1.186 | 4.185 | 0.684 | 680.0 | 316.9 |
| 100.0 | 1.0389 | 425.8 | 416.4 | 1.300 | 4.197 | 0.735 | 684.0 | 284.3 |
| 125.0 | 1.0599 | 531.1 | 521.6 | 1.573 | 4.235 | 0.863 | 688.9 | 224.5 |
| 150.0 | 1.0848 | 637.6 | 627.9 | 1.833 | 4.285 | 0.996 | 687.8 | 184.7 |
| 175.0 | 1.1142 | 745.5 | 735.5 | 2.080 | 4.353 | 1.145 | 681.5 | 156.8 |
| 200.0 | 1.1490 | 855.5 | 845.1 | 2.319 | 4.448 | 1.324 | 669.8 | 136.2 |
| 225.0 | 1.1909 | 968.3 | 957.6 | 2.551 | 4.586 | 1.551 | 652.3 | 120.4 |
| 250.0 | 1.2425 | 1085 | 1074 | 2.781 | 4.792 | 1.865 | 627.7 | 107.6 |
| 275.0 | 1.3091 | 1208 | 1197 | 3.011 | 5.120 | 2.347 | 594.0 | 96.52 |
| 300.0 | 1.4018 | 1343 | 1330 | 3.251 | 5.724 | 3.241 | 548.6 | 86.12 |
| 303.38 | | | | saturation | | | | |
| liquid | 1.4177 | 1363 | 1350 | 3.285 | 5.849 | 3.429 | 541.5 | 84.69 |
| vapour | 20.485 | 2741 | 2557 | 5.677 | 6.233 | 7.351 | 71.39 | 19.83 |
| 325.0 | 23.258 | 2854 | 2645 | 5.870 | 4.498 | 4.819 | 66.65 | 20.90 |
| 350.0 | 25.789 | 2955 | 2723 | 6.034 | 3.652 | 3.595 | 64.96 | 22.11 |
| 375.0 | 27.964 | 3040 | 2789 | 6.169 | 3.211 | 2.939 | 65.03 | 23.29 |
| 400.0 | 29.931 | 3117 | 2848 | 6.285 | 2.948 | 2.526 | 66.04 | 24.44 |
| 450.0 | 33.495 | 3256 | 2955 | 6.485 | 2.665 | 2.025 | 69.66 | 26.68 |
| 500.0 | 36.767 | 3386 | 3055 | 6.658 | 2.530 | 1.725 | 74.54 | 28.85 |
| 550.0 | 39.863 | 3510 | 3152 | 6.814 | 2.462 | 1.521 | 80.17 | 30.96 |
| 600.0 | 42.843 | 3633 | 3247 | 6.958 | 2.430 | 1.370 | 86.27 | 33.02 |
| 650.0 | 45.742 | 3754 | 3342 | 7.093 | 2.418 | 1.253 | 92.66 | 35.04 |
| 700.0 | 48.580 | 3875 | 3437 | 7.221 | 2.419 | 1.158 | 99.19 | 37.01 |
| 750.0 | 51.373 | 3996 | 3534 | 7.342 | 2.429 | 1.080 | 105.8 | 38.93 |
| 800.0 | 54.131 | 4118 | 3631 | 7.458 | 2.444 | 1.013 | 112.5 | 40.82 |
| 850.0 | 56.861 | 4240 | 3729 | 7.570 | 2.464 | 0.956 | 119.2 | 42.67 |
| 900.0 | 59.568 | 4364 | 3828 | 7.678 | 2.486 | 0.906 | 125.9 | 44.49 |
| 950.0 | 62.256 | 4489 | 3929 | 7.782 | 2.509 | 0.861 | 132.6 | 46.27 |
| 1000.0 | 64.930 | 4615 | 4031 | 7.883 | 2.534 | 0.821 | 139.3 | 48.02 |

Table 1 (cont.)

$$p = 10.0\,\text{MPa} = 100\ \text{bar}$$

| $t$ | $v$ | $h$ | $u$ | $s$ | $c_p$ | $\alpha_p$ | $\lambda$ | $\eta$ |
|---|---|---|---|---|---|---|---|---|
| °C | dm³/kg | kJ/kg | kJ/kg | kJ/kg K | kJ/kg K | $10^{-3}$/K | mW/K m | $\mu$Pa s |
| 0.0 | 0.9952 | 10.1 | 0.1 | 0.0004 | 4.177 | −0.040 | 566.5 | 1768 |
| 5.0 | 0.9952 | 30.9 | 21.0 | 0.076 | 4.156 | 0.042 | 575.6 | 1503 |
| 10.0 | 0.9956 | 51.7 | 41.7 | 0.150 | 4.150 | 0.111 | 584.9 | 1296 |
| 15.0 | 0.9963 | 72.4 | 62.5 | 0.223 | 4.151 | 0.171 | 594.0 | 1131 |
| 20.0 | 0.9973 | 93.2 | 83.2 | 0.294 | 4.153 | 0.223 | 603.0 | 998.1 |
| 25.0 | 0.9985 | 114.0 | 104.0 | 0.364 | 4.156 | 0.270 | 611.7 | 888.4 |
| 30.0 | 1.0000 | 134.8 | 124.8 | 0.433 | 4.157 | 0.313 | 619.9 | 796.9 |
| 35.0 | 1.0016 | 155.5 | 145.5 | 0.501 | 4.158 | 0.352 | 627.8 | 719.8 |
| 40.0 | 1.0035 | 176.3 | 166.3 | 0.568 | 4.159 | 0.388 | 635.1 | 654.1 |
| 45.0 | 1.0055 | 197.1 | 187.1 | 0.634 | 4.159 | 0.423 | 641.9 | 597.7 |
| 50.0 | 1.0078 | 217.9 | 207.8 | 0.699 | 4.160 | 0.456 | 648.2 | 548.9 |
| 60.0 | 1.0127 | 259.5 | 249.4 | 0.826 | 4.161 | 0.517 | 659.1 | 468.8 |
| 70.0 | 1.0182 | 301.2 | 291.0 | 0.949 | 4.166 | 0.575 | 668.0 | 406.6 |
| 80.0 | 1.0244 | 342.9 | 332.6 | 1.069 | 4.173 | 0.630 | 675.0 | 357.1 |
| 90.0 | 1.0311 | 384.6 | 374.3 | 1.185 | 4.183 | 0.682 | 680.5 | 317.2 |
| 100.0 | 1.0384 | 426.5 | 416.1 | 1.299 | 4.195 | 0.734 | 684.5 | 284.5 |
| 125.0 | 1.0593 | 531.8 | 521.2 | 1.572 | 4.233 | 0.861 | 689.5 | 224.7 |
| 150.0 | 1.0842 | 638.2 | 627.4 | 1.832 | 4.282 | 0.993 | 688.5 | 184.9 |
| 175.0 | 1.1134 | 746.1 | 735.0 | 2.079 | 4.349 | 1.141 | 682.2 | 157.0 |
| 200.0 | 1.1481 | 855.9 | 844.4 | 2.318 | 4.442 | 1.317 | 670.7 | 136.5 |
| 225.0 | 1.1897 | 968.6 | 956.7 | 2.550 | 4.578 | 1.541 | 653.3 | 120.7 |
| 250.0 | 1.2409 | 1085 | 1072 | 2.778 | 4.780 | 1.848 | 629.0 | 107.8 |
| 275.0 | 1.3065 | 1208 | 1195 | 3.008 | 5.099 | 2.316 | 595.7 | 96.83 |
| 300.0 | 1.3975 | 1342 | 1328 | 3.247 | 5.676 | 3.167 | 550.9 | 86.52 |
| 311.03 | saturation | | | | | | | |
| liquid | 1.4522 | 1407 | 1392 | 3.359 | 6.124 | 3.847 | 527.0 | 81.84 |
| vapour | 18.025 | 2724 | 2544 | 5.614 | 6.897 | 8.263 | 76.33 | 20.27 |
| 325.0 | 19.855 | 2808 | 2609 | 5.755 | 5.284 | 5.892 | 71.55 | 20.94 |
| 350.0 | 22.416 | 2922 | 2698 | 5.943 | 4.028 | 4.082 | 68.11 | 22.15 |
| 375.0 | 24.532 | 3014 | 2769 | 6.088 | 3.436 | 3.221 | 67.35 | 23.33 |
| 400.0 | 26.408 | 3096 | 2832 | 6.211 | 3.100 | 2.711 | 67.89 | 24.48 |
| 450.0 | 29.752 | 3241 | 2943 | 6.419 | 2.749 | 2.122 | 70.99 | 26.73 |
| 500.0 | 32.784 | 3373 | 3046 | 6.597 | 2.584 | 1.785 | 75.61 | 28.91 |
| 550.0 | 35.632 | 3500 | 3144 | 6.756 | 2.500 | 1.561 | 81.11 | 31.02 |
| 600.0 | 38.361 | 3624 | 3241 | 6.902 | 2.458 | 1.398 | 87.13 | 33.09 |
| 650.0 | 41.006 | 3747 | 3337 | 7.038 | 2.440 | 1.274 | 93.47 | 35.10 |
| 700.0 | 43.590 | 3868 | 3433 | 7.167 | 2.437 | 1.174 | 99.97 | 37.07 |
| 750.0 | 46.128 | 3990 | 3529 | 7.289 | 2.444 | 1.092 | 106.5 | 38.99 |
| 800.0 | 48.630 | 4113 | 3627 | 7.406 | 2.456 | 1.023 | 113.2 | 40.88 |
| 850.0 | 51.104 | 4236 | 3725 | 7.518 | 2.474 | 0.963 | 119.8 | 42.73 |
| 900.0 | 53.555 | 4360 | 3825 | 7.627 | 2.494 | 0.912 | 126.4 | 44.54 |
| 950.0 | 55.987 | 4486 | 3926 | 7.731 | 2.516 | 0.866 | 133.1 | 46.32 |
| 1000.0 | 58.404 | 4612 | 4028 | 7.832 | 2.540 | 0.825 | 139.7 | 48.07 |

Table 1 (cont.)

$$p = 11.0\,\text{MPa} = 110\ \text{bar}$$

| $t$ | $v$ | $h$ | $u$ | $s$ | $c_p$ | $\alpha_p$ | $\lambda$ | $\eta$ |
|---|---|---|---|---|---|---|---|---|
| °C | dm³/kg | kJ/kg | kJ/kg | kJ/kg K | kJ/kg K | $10^{-3}$/K | mW/K m | $\mu$Pa s |
| 0.0 | 0.9947 | 11.1 | 0.2 | 0.0005 | 4.172 | −0.036 | 567.1 | 1766 |
| 5.0 | 0.9947 | 31.9 | 21.0 | 0.076 | 4.152 | 0.045 | 576.2 | 1501 |
| 10.0 | 0.9951 | 52.6 | 41.7 | 0.150 | 4.147 | 0.113 | 585.3 | 1295 |
| 15.0 | 0.9959 | 73.4 | 62.4 | 0.222 | 4.148 | 0.172 | 594.5 | 1130 |
| 20.0 | 0.9969 | 94.1 | 83.2 | 0.294 | 4.150 | 0.224 | 603.4 | 997.7 |
| 25.0 | 0.9981 | 114.9 | 103.9 | 0.364 | 4.153 | 0.271 | 612.1 | 888.2 |
| 30.0 | 0.9995 | 135.7 | 124.7 | 0.433 | 4.155 | 0.313 | 620.4 | 796.9 |
| 35.0 | 1.0012 | 156.4 | 145.4 | 0.501 | 4.156 | 0.352 | 628.2 | 719.8 |
| 40.0 | 1.0031 | 177.2 | 166.2 | 0.568 | 4.157 | 0.389 | 635.5 | 654.2 |
| 45.0 | 1.0051 | 198.0 | 186.9 | 0.634 | 4.157 | 0.423 | 642.4 | 597.9 |
| 50.0 | 1.0073 | 218.8 | 207.7 | 0.699 | 4.157 | 0.456 | 648.6 | 549.0 |
| 60.0 | 1.0122 | 260.4 | 249.2 | 0.825 | 4.159 | 0.517 | 659.6 | 469.1 |
| 70.0 | 1.0178 | 302.0 | 290.8 | 0.948 | 4.164 | 0.574 | 668.5 | 406.8 |
| 80.0 | 1.0239 | 343.6 | 332.4 | 1.068 | 4.171 | 0.629 | 675.6 | 357.4 |
| 90.0 | 1.0306 | 385.4 | 374.1 | 1.185 | 4.181 | 0.681 | 681.0 | 317.5 |
| 100.0 | 1.0380 | 427.3 | 415.9 | 1.298 | 4.193 | 0.732 | 685.1 | 284.8 |
| 125.0 | 1.0588 | 532.5 | 520.9 | 1.572 | 4.230 | 0.858 | 690.1 | 225.0 |
| 150.0 | 1.0835 | 638.9 | 627.0 | 1.831 | 4.279 | 0.990 | 689.2 | 185.2 |
| 175.0 | 1.1126 | 746.6 | 734.4 | 2.078 | 4.345 | 1.136 | 683.0 | 157.3 |
| 200.0 | 1.1471 | 856.3 | 843.7 | 2.316 | 4.437 | 1.310 | 671.6 | 136.7 |
| 225.0 | 1.1884 | 968.8 | 955.8 | 2.548 | 4.571 | 1.531 | 654.3 | 120.9 |
| 250.0 | 1.2392 | 1085 | 1071 | 2.776 | 4.768 | 1.832 | 630.2 | 108.1 |
| 275.0 | 1.3041 | 1208 | 1193 | 3.005 | 5.078 | 2.287 | 597.3 | 97.14 |
| 300.0 | 1.3932 | 1341 | 1325 | 3.243 | 5.630 | 3.098 | 553.2 | 86.92 |
| 318.11 | | | | saturation | | | | |
| liquid | 1.4881 | 1449 | 1433 | 3.429 | 6.443 | 4.338 | 513.1 | 79.18 |
| vapour | 15.986 | 2705 | 2529 | 5.553 | 7.662 | 9.339 | 81.90 | 20.72 |
| 325.0 | 16.934 | 2753 | 2567 | 5.634 | 6.456 | 7.545 | 78.16 | 21.03 |
| 350.0 | 19.604 | 2886 | 2670 | 5.851 | 4.494 | 4.699 | 71.85 | 22.21 |
| 375.0 | 21.698 | 2987 | 2748 | 6.010 | 3.697 | 3.552 | 69.98 | 23.39 |
| 400.0 | 23.510 | 3073 | 2815 | 6.141 | 3.269 | 2.918 | 69.92 | 24.54 |
| 450.0 | 26.683 | 3225 | 2931 | 6.358 | 2.839 | 2.227 | 72.41 | 26.79 |
| 500.0 | 29.521 | 3361 | 3036 | 6.541 | 2.641 | 1.848 | 76.74 | 28.97 |
| 550.0 | 32.168 | 3490 | 3136 | 6.703 | 2.540 | 1.602 | 82.08 | 31.09 |
| 600.0 | 34.692 | 3616 | 3234 | 6.851 | 2.487 | 1.428 | 88.03 | 33.15 |
| 650.0 | 37.130 | 3739 | 3331 | 6.988 | 2.463 | 1.295 | 94.31 | 35.17 |
| 700.0 | 39.506 | 3862 | 3428 | 7.118 | 2.455 | 1.190 | 100.8 | 37.13 |
| 750.0 | 41.836 | 3985 | 3525 | 7.241 | 2.458 | 1.104 | 107.3 | 39.06 |
| 800.0 | 44.129 | 4108 | 3623 | 7.359 | 2.469 | 1.032 | 113.9 | 40.94 |
| 850.0 | 46.394 | 4232 | 3722 | 7.471 | 2.484 | 0.971 | 120.4 | 42.79 |
| 900.0 | 48.635 | 4357 | 3822 | 7.580 | 2.503 | 0.918 | 127.0 | 44.60 |
| 950.0 | 50.858 | 4482 | 3923 | 7.685 | 2.524 | 0.871 | 133.6 | 46.38 |
| 1000.0 | 53.066 | 4609 | 4025 | 7.786 | 2.546 | 0.829 | 140.2 | 48.12 |

Table 1 (cont.) $p = 12.0\,\text{MPa} = 120\ \text{bar}$

| $t$ | $v$ | $h$ | $u$ | $s$ | $c_p$ | $\alpha_p$ | $\lambda$ | $\eta$ |
|---|---|---|---|---|---|---|---|---|
| °C | dm³/kg | kJ/kg | kJ/kg | kJ/kg K | kJ/kg K | $10^{-3}$/K | mW/K m | $\mu$Pa s |
| 0.0 | 0.9942 | 12.1 | 0.2 | 0.0005 | 4.167 | −0.033 | 567.6 | 1764 |
| 5.0 | 0.9943 | 32.9 | 21.0 | 0.076 | 4.148 | 0.048 | 576.7 | 1500 |
| 10.0 | 0.9947 | 53.6 | 41.7 | 0.150 | 4.143 | 0.116 | 585.8 | 1294 |
| 15.0 | 0.9954 | 74.3 | 62.4 | 0.222 | 4.144 | 0.174 | 595.0 | 1130 |
| 20.0 | 0.9964 | 95.1 | 83.1 | 0.294 | 4.147 | 0.226 | 603.9 | 997.4 |
| 25.0 | 0.9977 | 115.8 | 103.8 | 0.364 | 4.150 | 0.272 | 612.6 | 888.0 |
| 30.0 | 0.9991 | 136.6 | 124.6 | 0.433 | 4.152 | 0.314 | 620.8 | 796.8 |
| 35.0 | 1.0008 | 157.3 | 145.3 | 0.501 | 4.154 | 0.353 | 628.7 | 719.9 |
| 40.0 | 1.0026 | 178.1 | 166.1 | 0.568 | 4.154 | 0.389 | 636.0 | 654.4 |
| 45.0 | 1.0047 | 198.9 | 186.8 | 0.633 | 4.155 | 0.423 | 642.8 | 598.0 |
| 50.0 | 1.0069 | 219.6 | 207.6 | 0.698 | 4.155 | 0.456 | 649.1 | 549.2 |
| 60.0 | 1.0118 | 261.2 | 249.1 | 0.825 | 4.157 | 0.516 | 660.0 | 469.3 |
| 70.0 | 1.0173 | 302.8 | 290.6 | 0.948 | 4.162 | 0.573 | 669.0 | 407.1 |
| 80.0 | 1.0235 | 344.4 | 332.2 | 1.068 | 4.169 | 0.628 | 676.1 | 357.7 |
| 90.0 | 1.0302 | 386.2 | 373.8 | 1.184 | 4.179 | 0.680 | 681.5 | 317.7 |
| 100.0 | 1.0375 | 428.0 | 415.6 | 1.298 | 4.190 | 0.731 | 685.6 | 285.1 |
| 125.0 | 1.0582 | 533.2 | 520.5 | 1.571 | 4.228 | 0.856 | 690.7 | 225.3 |
| 150.0 | 1.0829 | 639.5 | 626.5 | 1.829 | 4.276 | 0.987 | 689.8 | 185.4 |
| 175.0 | 1.1119 | 747.2 | 733.8 | 2.077 | 4.341 | 1.132 | 683.7 | 157.5 |
| 200.0 | 1.1462 | 856.8 | 843.0 | 2.315 | 4.432 | 1.304 | 672.4 | 137.0 |
| 225.0 | 1.1872 | 969.1 | 954.9 | 2.546 | 4.563 | 1.521 | 655.3 | 121.2 |
| 250.0 | 1.2375 | 1085 | 1070 | 2.774 | 4.757 | 1.816 | 631.4 | 108.4 |
| 275.0 | 1.3016 | 1207 | 1192 | 3.002 | 5.059 | 2.258 | 598.8 | 97.44 |
| 300.0 | 1.3892 | 1340 | 1323 | 3.238 | 5.586 | 3.034 | 555.4 | 87.30 |
| 324.71 | saturation | | | | | | | |
| liquid | 1.5259 | 1490 | 1472 | 3.495 | 6.820 | 4.926 | 499.9 | 76.65 |
| vapour | 14.262 | 2684 | 2513 | 5.492 | 8.558 | 10.63 | 88.27 | 21.18 |
| 325.0 | 14.306 | 2687 | 2515 | 5.496 | 8.473 | 10.50 | 88.00 | 21.19 |
| 350.0 | 17.203 | 2846 | 2639 | 5.758 | 5.092 | 5.507 | 76.44 | 22.31 |
| 375.0 | 19.309 | 2958 | 2726 | 5.934 | 4.001 | 3.943 | 73.00 | 23.46 |
| 400.0 | 21.080 | 3050 | 2797 | 6.074 | 3.457 | 3.152 | 72.18 | 24.61 |
| 450.0 | 24.120 | 3208 | 2919 | 6.300 | 2.934 | 2.339 | 73.93 | 26.86 |
| 500.0 | 26.800 | 3348 | 3027 | 6.488 | 2.699 | 1.913 | 77.92 | 29.04 |
| 550.0 | 29.281 | 3480 | 3128 | 6.653 | 2.580 | 1.645 | 83.09 | 31.16 |
| 600.0 | 31.634 | 3607 | 3228 | 6.803 | 2.517 | 1.457 | 88.95 | 33.22 |
| 650.0 | 33.901 | 3732 | 3325 | 6.942 | 2.486 | 1.317 | 95.18 | 35.24 |
| 700.0 | 36.104 | 3856 | 3423 | 7.073 | 2.473 | 1.206 | 101.6 | 37.20 |
| 750.0 | 38.260 | 3980 | 3521 | 7.197 | 2.473 | 1.117 | 108.1 | 39.13 |
| 800.0 | 40.379 | 4103 | 3619 | 7.315 | 2.481 | 1.042 | 114.6 | 41.01 |
| 850.0 | 42.469 | 4228 | 3718 | 7.428 | 2.494 | 0.979 | 121.1 | 42.85 |
| 900.0 | 44.536 | 4353 | 3818 | 7.537 | 2.512 | 0.924 | 127.6 | 44.66 |
| 950.0 | 46.585 | 4479 | 3920 | 7.642 | 2.531 | 0.876 | 134.1 | 46.43 |
| 1000.0 | 48.618 | 4606 | 4023 | 7.744 | 2.553 | 0.833 | 140.6 | 48.17 |

Table 1 (cont.)

$$p = 13.0\,\text{MPa} = 130\;\text{bar}$$

| $t$ | $v$ | $h$ | $u$ | $s$ | $c_p$ | $\alpha_p$ | $\lambda$ | $\eta$ |
|---|---|---|---|---|---|---|---|---|
| °C | dm³/kg | kJ/kg | kJ/kg | kJ/kg K | kJ/kg K | $10^{-3}$/K | mW/K m | $\mu$Pa s |
| 0.0 | 0.9937 | 13.1 | 0.2 | 0.0005 | 4.162 | −0.029 | 568.2 | 1762 |
| 5.0 | 0.9938 | 33.9 | 20.9 | 0.076 | 4.144 | 0.051 | 577.2 | 1498 |
| 10.0 | 0.9942 | 54.6 | 41.6 | 0.150 | 4.140 | 0.118 | 586.3 | 1293 |
| 15.0 | 0.9950 | 75.3 | 62.3 | 0.222 | 4.141 | 0.176 | 595.4 | 1129 |
| 20.0 | 0.9960 | 96.0 | 83.0 | 0.293 | 4.145 | 0.227 | 604.4 | 997.0 |
| 25.0 | 0.9972 | 116.7 | 103.8 | 0.363 | 4.148 | 0.273 | 613.0 | 887.8 |
| 30.0 | 0.9987 | 137.5 | 124.5 | 0.432 | 4.150 | 0.315 | 621.3 | 796.8 |
| 35.0 | 1.0004 | 158.2 | 145.2 | 0.500 | 4.151 | 0.353 | 629.1 | 719.9 |
| 40.0 | 1.0022 | 179.0 | 165.9 | 0.567 | 4.152 | 0.389 | 636.5 | 654.5 |
| 45.0 | 1.0043 | 199.7 | 186.7 | 0.633 | 4.153 | 0.423 | 643.3 | 598.2 |
| 50.0 | 1.0065 | 220.5 | 207.4 | 0.698 | 4.153 | 0.455 | 649.5 | 549.4 |
| 60.0 | 1.0114 | 262.0 | 248.9 | 0.824 | 4.155 | 0.516 | 660.5 | 469.5 |
| 70.0 | 1.0169 | 303.6 | 290.4 | 0.947 | 4.160 | 0.573 | 669.5 | 407.3 |
| 80.0 | 1.0230 | 345.2 | 331.9 | 1.067 | 4.167 | 0.627 | 676.6 | 357.9 |
| 90.0 | 1.0297 | 387.0 | 373.6 | 1.183 | 4.176 | 0.678 | 682.1 | 318.0 |
| 100.0 | 1.0370 | 428.8 | 415.3 | 1.297 | 4.188 | 0.729 | 686.2 | 285.3 |
| 125.0 | 1.0577 | 533.9 | 520.2 | 1.570 | 4.225 | 0.854 | 691.3 | 225.5 |
| 150.0 | 1.0822 | 640.1 | 626.1 | 1.828 | 4.273 | 0.983 | 690.5 | 185.7 |
| 175.0 | 1.1111 | 747.7 | 733.3 | 2.075 | 4.338 | 1.127 | 684.5 | 157.7 |
| 200.0 | 1.1452 | 857.2 | 842.3 | 2.313 | 4.427 | 1.297 | 673.3 | 137.2 |
| 225.0 | 1.1860 | 969.4 | 954.0 | 2.544 | 4.556 | 1.511 | 656.3 | 121.4 |
| 250.0 | 1.2359 | 1085 | 1069 | 2.772 | 4.746 | 1.801 | 632.6 | 108.6 |
| 275.0 | 1.2993 | 1207 | 1190 | 2.999 | 5.039 | 2.231 | 600.4 | 97.75 |
| 300.0 | 1.3852 | 1339 | 1321 | 3.234 | 5.546 | 2.973 | 557.6 | 87.67 |
| 325.0 | 1.5193 | 1489 | 1470 | 3.491 | 6.703 | 4.741 | 502.4 | 77.09 |
| 330.89 | | | | saturation | | | | |
| liquid | 1.5662 | 1530 | 1510 | 3.559 | 7.274 | 5.647 | 487.4 | 74.22 |
| vapour | 12.779 | 2661 | 2495 | 5.432 | 9.629 | 12.19 | 95.65 | 21.68 |
| 350.0 | 15.103 | 2801 | 2605 | 5.661 | 5.892 | 6.615 | 82.24 | 22.44 |
| 375.0 | 17.259 | 2927 | 2702 | 5.858 | 4.360 | 4.413 | 76.49 | 23.56 |
| 400.0 | 19.009 | 3026 | 2779 | 6.008 | 3.668 | 3.417 | 74.69 | 24.70 |
| 450.0 | 21.945 | 3191 | 2906 | 6.245 | 3.035 | 2.460 | 75.56 | 26.94 |
| 500.0 | 24.495 | 3335 | 3017 | 6.438 | 2.761 | 1.982 | 79.16 | 29.12 |
| 550.0 | 26.836 | 3469 | 3121 | 6.606 | 2.622 | 1.689 | 84.15 | 31.23 |
| 600.0 | 29.046 | 3598 | 3221 | 6.758 | 2.548 | 1.488 | 89.91 | 33.30 |
| 650.0 | 31.168 | 3725 | 3320 | 6.899 | 2.509 | 1.338 | 96.08 | 35.31 |
| 700.0 | 33.225 | 3850 | 3418 | 7.031 | 2.492 | 1.222 | 102.4 | 37.27 |
| 750.0 | 35.234 | 3974 | 3516 | 7.155 | 2.488 | 1.129 | 108.9 | 39.19 |
| 800.0 | 37.206 | 4099 | 3615 | 7.274 | 2.493 | 1.052 | 115.4 | 41.07 |
| 850.0 | 39.148 | 4224 | 3715 | 7.388 | 2.505 | 0.986 | 121.8 | 42.91 |
| 900.0 | 41.068 | 4349 | 3815 | 7.497 | 2.520 | 0.930 | 128.3 | 44.72 |
| 950.0 | 42.969 | 4476 | 3917 | 7.603 | 2.539 | 0.881 | 134.7 | 46.49 |
| 1000.0 | 44.854 | 4603 | 4020 | 7.705 | 2.559 | 0.838 | 141.1 | 48.22 |

Table 1 (cont.)

$$p = 14.0\,\text{MPa} = 140\ \text{bar}$$

| $t$ | $v$ | $h$ | $u$ | $s$ | $c_p$ | $\alpha_p$ | $\lambda$ | $\eta$ |
|---|---|---|---|---|---|---|---|---|
| °C | dm³/kg | kJ/kg | kJ/kg | kJ/kg K | kJ/kg K | $10^{-3}$/K | mW/K m | $\mu$Pa s |
| 0.0 | 0.9933 | 14.1 | 0.2 | 0.0006 | 4.157 | −0.025 | 568.7 | 1759 |
| 5.0 | 0.9933 | 34.8 | 20.9 | 0.076 | 4.140 | 0.054 | 577.7 | 1497 |
| 10.0 | 0.9938 | 55.5 | 41.6 | 0.150 | 4.136 | 0.120 | 586.8 | 1292 |
| 15.0 | 0.9945 | 76.2 | 62.3 | 0.222 | 4.138 | 0.178 | 595.9 | 1129 |
| 20.0 | 0.9955 | 96.9 | 83.0 | 0.293 | 4.142 | 0.229 | 604.8 | 996.7 |
| 25.0 | 0.9968 | 117.6 | 103.7 | 0.363 | 4.145 | 0.274 | 613.5 | 887.7 |
| 30.0 | 0.9983 | 138.4 | 124.4 | 0.432 | 4.148 | 0.315 | 621.8 | 796.7 |
| 35.0 | 0.9999 | 159.1 | 145.1 | 0.500 | 4.149 | 0.354 | 629.6 | 720.0 |
| 40.0 | 1.0018 | 179.9 | 165.8 | 0.567 | 4.150 | 0.390 | 636.9 | 654.6 |
| 45.0 | 1.0038 | 200.6 | 186.6 | 0.633 | 4.150 | 0.423 | 643.7 | 598.3 |
| 50.0 | 1.0060 | 221.4 | 207.3 | 0.697 | 4.151 | 0.455 | 650.0 | 549.6 |
| 60.0 | 1.0109 | 262.9 | 248.7 | 0.824 | 4.153 | 0.516 | 661.0 | 469.8 |
| 70.0 | 1.0164 | 304.4 | 290.2 | 0.947 | 4.158 | 0.572 | 669.9 | 407.6 |
| 80.0 | 1.0226 | 346.0 | 331.7 | 1.066 | 4.165 | 0.626 | 677.1 | 358.2 |
| 90.0 | 1.0292 | 387.7 | 373.3 | 1.183 | 4.174 | 0.677 | 682.6 | 318.3 |
| 100.0 | 1.0365 | 429.5 | 415.0 | 1.296 | 4.186 | 0.727 | 686.7 | 285.6 |
| 125.0 | 1.0571 | 534.6 | 519.8 | 1.569 | 4.223 | 0.851 | 691.9 | 225.8 |
| 150.0 | 1.0816 | 640.8 | 625.6 | 1.827 | 4.271 | 0.980 | 691.2 | 185.9 |
| 175.0 | 1.1104 | 748.3 | 732.7 | 2.074 | 4.334 | 1.123 | 685.2 | 158.0 |
| 200.0 | 1.1443 | 857.7 | 841.7 | 2.312 | 4.422 | 1.291 | 674.1 | 137.4 |
| 225.0 | 1.1848 | 969.7 | 953.1 | 2.542 | 4.549 | 1.502 | 657.3 | 121.6 |
| 250.0 | 1.2343 | 1085 | 1068 | 2.769 | 4.735 | 1.786 | 633.8 | 108.9 |
| 275.0 | 1.2969 | 1207 | 1189 | 2.997 | 5.021 | 2.205 | 601.9 | 98.04 |
| 300.0 | 1.3814 | 1338 | 1318 | 3.230 | 5.507 | 2.916 | 559.7 | 88.04 |
| 325.0 | 1.5111 | 1487 | 1465 | 3.484 | 6.577 | 4.543 | 505.6 | 77.65 |
| 336.70 | | | | saturation | | | | |
| liquid | 1.6096 | 1570 | 1547 | 3.622 | 7.836 | 6.552 | 475.4 | 71.85 |
| vapour | 11.485 | 2637 | 2476 | 5.371 | 10.94 | 14.14 | 104.3 | 22.21 |
| 350.0 | 13.218 | 2751 | 2566 | 5.557 | 7.033 | 8.239 | 89.94 | 22.64 |
| 375.0 | 15.472 | 2893 | 2677 | 5.781 | 4.793 | 4.989 | 80.61 | 23.69 |
| 400.0 | 17.219 | 3001 | 2760 | 5.944 | 3.906 | 3.720 | 77.50 | 24.80 |
| 450.0 | 20.075 | 3174 | 2893 | 6.192 | 3.144 | 2.589 | 77.31 | 27.03 |
| 500.0 | 22.517 | 3322 | 3007 | 6.390 | 2.824 | 2.054 | 80.47 | 29.20 |
| 550.0 | 24.739 | 3459 | 3113 | 6.562 | 2.664 | 1.735 | 85.25 | 31.31 |
| 600.0 | 26.828 | 3590 | 3214 | 6.716 | 2.579 | 1.519 | 90.90 | 33.37 |
| 650.0 | 28.825 | 3717 | 3314 | 6.859 | 2.533 | 1.361 | 97.00 | 35.38 |
| 700.0 | 30.757 | 3843 | 3413 | 6.991 | 2.510 | 1.239 | 103.3 | 37.35 |
| 750.0 | 32.640 | 3969 | 3512 | 7.117 | 2.503 | 1.142 | 109.7 | 39.26 |
| 800.0 | 34.486 | 4094 | 3611 | 7.236 | 2.506 | 1.061 | 116.1 | 41.14 |
| 850.0 | 36.303 | 4219 | 3711 | 7.351 | 2.515 | 0.994 | 122.5 | 42.98 |
| 900.0 | 38.096 | 4345 | 3812 | 7.461 | 2.529 | 0.936 | 128.9 | 44.78 |
| 950.0 | 39.870 | 4472 | 3914 | 7.566 | 2.547 | 0.886 | 135.3 | 46.54 |
| 1000.0 | 41.628 | 4600 | 4017 | 7.669 | 2.566 | 0.842 | 141.6 | 48.27 |

Table 1 (cont.)

$$p = 15.0\,\text{MPa} = 150\ \text{bar}$$

| $t$ | $v$ | $h$ | $u$ | $s$ | $c_p$ | $\alpha_p$ | $\lambda$ | $\eta$ |
|---|---|---|---|---|---|---|---|---|
| °C | dm³/kg | kJ/kg | kJ/kg | kJ/kg K | kJ/kg K | $10^{-3}$/K | mW/K m | $\mu$Pa s |
| 0.0 | 0.9928 | 15.1 | 0.2 | 0.0006 | 4.153 | −0.021 | 569.3 | 1757 |
| 5.0 | 0.9929 | 35.8 | 20.9 | 0.076 | 4.136 | 0.057 | 578.2 | 1495 |
| 10.0 | 0.9933 | 56.5 | 41.6 | 0.149 | 4.133 | 0.123 | 587.3 | 1291 |
| 15.0 | 0.9941 | 77.2 | 62.3 | 0.222 | 4.135 | 0.180 | 596.4 | 1128 |
| 20.0 | 0.9951 | 97.8 | 82.9 | 0.293 | 4.139 | 0.230 | 605.3 | 996.4 |
| 25.0 | 0.9963 | 118.5 | 103.6 | 0.363 | 4.142 | 0.275 | 613.9 | 887.5 |
| 30.0 | 0.9978 | 139.3 | 124.3 | 0.432 | 4.145 | 0.316 | 622.2 | 796.7 |
| 35.0 | 0.9995 | 160.0 | 145.0 | 0.500 | 4.147 | 0.354 | 630.0 | 720.0 |
| 40.0 | 1.0014 | 180.7 | 165.7 | 0.566 | 4.148 | 0.390 | 637.4 | 654.7 |
| 45.0 | 1.0034 | 201.5 | 186.4 | 0.632 | 4.148 | 0.423 | 644.2 | 598.5 |
| 50.0 | 1.0056 | 222.2 | 207.1 | 0.697 | 4.149 | 0.455 | 650.5 | 549.8 |
| 60.0 | 1.0105 | 263.7 | 248.6 | 0.823 | 4.151 | 0.515 | 661.5 | 470.0 |
| 70.0 | 1.0160 | 305.2 | 290.0 | 0.946 | 4.156 | 0.571 | 670.4 | 407.8 |
| 80.0 | 1.0221 | 346.8 | 331.5 | 1.066 | 4.163 | 0.625 | 677.6 | 358.5 |
| 90.0 | 1.0288 | 388.5 | 373.1 | 1.182 | 4.172 | 0.676 | 683.1 | 318.6 |
| 100.0 | 1.0360 | 430.3 | 414.7 | 1.295 | 4.184 | 0.726 | 687.2 | 285.9 |
| 125.0 | 1.0566 | 535.3 | 519.5 | 1.568 | 4.221 | 0.849 | 692.5 | 226.0 |
| 150.0 | 1.0810 | 641.4 | 625.2 | 1.826 | 4.268 | 0.977 | 691.8 | 186.2 |
| 175.0 | 1.1096 | 748.8 | 732.2 | 2.073 | 4.330 | 1.118 | 686.0 | 158.2 |
| 200.0 | 1.1434 | 858.1 | 841.0 | 2.310 | 4.417 | 1.285 | 674.9 | 137.7 |
| 225.0 | 1.1836 | 970.0 | 952.3 | 2.541 | 4.542 | 1.493 | 658.3 | 121.9 |
| 250.0 | 1.2327 | 1085 | 1067 | 2.767 | 4.724 | 1.772 | 635.0 | 109.2 |
| 275.0 | 1.2946 | 1207 | 1187 | 2.994 | 5.003 | 2.179 | 603.4 | 98.34 |
| 300.0 | 1.3777 | 1337 | 1316 | 3.226 | 5.470 | 2.862 | 561.7 | 88.40 |
| 325.0 | 1.5033 | 1484 | 1462 | 3.477 | 6.465 | 4.367 | 508.6 | 78.18 |
| 342.19 | saturation | | | | | | | |
| liquid | 1.6571 | 1609 | 1584 | 3.684 | 8.551 | 7.725 | 464.0 | 69.50 |
| vapour | 10.339 | 2610 | 2454 | 5.309 | 12.58 | 16.64 | 114.6 | 22.79 |
| 350.0 | 11.469 | 2691 | 2519 | 5.440 | 8.838 | 10.89 | 100.9 | 22.94 |
| 375.0 | 13.889 | 2857 | 2649 | 5.703 | 5.325 | 5.710 | 85.53 | 23.86 |
| 400.0 | 15.652 | 2974 | 2739 | 5.880 | 4.177 | 4.068 | 80.68 | 24.93 |
| 450.0 | 18.450 | 3156 | 2879 | 6.141 | 3.260 | 2.728 | 79.18 | 27.13 |
| 500.0 | 20.800 | 3309 | 2997 | 6.345 | 2.891 | 2.130 | 81.85 | 29.29 |
| 550.0 | 22.921 | 3448 | 3104 | 6.520 | 2.708 | 1.782 | 86.39 | 31.40 |
| 600.0 | 24.904 | 3581 | 3207 | 6.677 | 2.610 | 1.551 | 91.92 | 33.46 |
| 650.0 | 26.794 | 3710 | 3308 | 6.820 | 2.557 | 1.383 | 97.96 | 35.46 |
| 700.0 | 28.618 | 3837 | 3408 | 6.954 | 2.529 | 1.256 | 104.2 | 37.42 |
| 750.0 | 30.392 | 3963 | 3507 | 7.081 | 2.518 | 1.154 | 110.6 | 39.34 |
| 800.0 | 32.129 | 4089 | 3607 | 7.201 | 2.518 | 1.071 | 116.9 | 41.21 |
| 850.0 | 33.836 | 4215 | 3708 | 7.316 | 2.526 | 1.002 | 123.3 | 43.04 |
| 900.0 | 35.520 | 4342 | 3809 | 7.426 | 2.538 | 0.942 | 129.6 | 44.84 |
| 950.0 | 37.184 | 4469 | 3911 | 7.532 | 2.554 | 0.891 | 135.9 | 46.60 |
| 1000.0 | 38.833 | 4597 | 4015 | 7.635 | 2.573 | 0.846 | 142.1 | 48.33 |

Table 1 (cont.)

$$p = 16.0\,\text{MPa} = 160\ \text{bar}$$

| $t$ | $v$ | $h$ | $u$ | $s$ | $c_p$ | $\alpha_p$ | $\lambda$ | $\eta$ |
|------|------|------|------|------|------|------|------|------|
| °C | dm³/kg | kJ/kg | kJ/kg | kJ/kg K | kJ/kg K | $10^{-3}$/K | mW/K m | $\mu$Pa s |
| 0.0 | 0.9923 | 16.1 | 0.2 | 0.0006 | 4.148 | −0.018 | 569.8 | 1755 |
| 5.0 | 0.9924 | 36.8 | 20.9 | 0.076 | 4.132 | 0.060 | 578.7 | 1494 |
| 10.0 | 0.9929 | 57.5 | 41.6 | 0.149 | 4.129 | 0.125 | 587.8 | 1290 |
| 15.0 | 0.9936 | 78.1 | 62.2 | 0.222 | 4.132 | 0.181 | 596.8 | 1127 |
| 20.0 | 0.9946 | 98.8 | 82.9 | 0.293 | 4.136 | 0.231 | 605.7 | 996.1 |
| 25.0 | 0.9959 | 119.5 | 103.5 | 0.363 | 4.140 | 0.276 | 614.4 | 887.4 |
| 30.0 | 0.9974 | 140.2 | 124.2 | 0.432 | 4.143 | 0.317 | 622.7 | 796.6 |
| 35.0 | 0.9991 | 160.9 | 144.9 | 0.499 | 4.144 | 0.355 | 630.5 | 720.1 |
| 40.0 | 1.0009 | 181.6 | 165.6 | 0.566 | 4.146 | 0.390 | 637.8 | 654.8 |
| 45.0 | 1.0030 | 202.3 | 186.3 | 0.632 | 4.146 | 0.423 | 644.7 | 598.7 |
| 50.0 | 1.0052 | 223.1 | 207.0 | 0.696 | 4.147 | 0.455 | 650.9 | 550.0 |
| 60.0 | 1.0101 | 264.6 | 248.4 | 0.823 | 4.149 | 0.515 | 661.9 | 470.3 |
| 70.0 | 1.0156 | 306.1 | 289.8 | 0.946 | 4.154 | 0.571 | 670.9 | 408.1 |
| 80.0 | 1.0217 | 347.6 | 331.3 | 1.065 | 4.161 | 0.624 | 678.1 | 358.7 |
| 90.0 | 1.0283 | 389.3 | 372.8 | 1.181 | 4.170 | 0.675 | 683.6 | 318.8 |
| 100.0 | 1.0355 | 431.0 | 414.5 | 1.295 | 4.182 | 0.724 | 687.8 | 286.1 |
| 125.0 | 1.0561 | 536.0 | 519.1 | 1.567 | 4.218 | 0.847 | 693.1 | 226.3 |
| 150.0 | 1.0804 | 642.0 | 624.8 | 1.825 | 4.265 | 0.974 | 692.5 | 186.4 |
| 175.0 | 1.1089 | 749.4 | 731.6 | 2.072 | 4.327 | 1.114 | 686.7 | 158.5 |
| 200.0 | 1.1425 | 858.6 | 840.3 | 2.309 | 4.412 | 1.278 | 675.8 | 137.9 |
| 225.0 | 1.1825 | 970.3 | 951.4 | 2.539 | 4.535 | 1.484 | 659.2 | 122.1 |
| 250.0 | 1.2311 | 1085 | 1066 | 2.765 | 4.714 | 1.757 | 636.1 | 109.4 |
| 275.0 | 1.2924 | 1206 | 1186 | 2.991 | 4.985 | 2.154 | 604.9 | 98.63 |
| 300.0 | 1.3740 | 1336 | 1314 | 3.222 | 5.435 | 2.811 | 563.8 | 88.76 |
| 325.0 | 1.4960 | 1482 | 1458 | 3.471 | 6.364 | 4.209 | 511.6 | 78.69 |
| 347.39 | | | | saturation | | | | |
| liquid | 1.7099 | 1649 | 1622 | 3.745 | 9.498 | 9.305 | 453.1 | 67.13 |
| vapour | 9.3105 | 2580 | 2431 | 5.245 | 14.72 | 19.94 | 127.1 | 23.44 |
| 350.0 | 9.7553 | 2615 | 2459 | 5.301 | 12.32 | 16.19 | 118.8 | 23.41 |
| 375.0 | 12.467 | 2818 | 2619 | 5.622 | 5.998 | 6.639 | 91.55 | 24.07 |
| 400.0 | 14.265 | 2946 | 2718 | 5.816 | 4.487 | 4.473 | 84.29 | 25.08 |
| 450.0 | 17.023 | 3138 | 2866 | 6.091 | 3.385 | 2.879 | 81.20 | 27.24 |
| 500.0 | 19.296 | 3295 | 2986 | 6.302 | 2.960 | 2.209 | 83.30 | 29.38 |
| 550.0 | 21.330 | 3438 | 3096 | 6.480 | 2.754 | 1.830 | 87.58 | 31.49 |
| 600.0 | 23.221 | 3572 | 3201 | 6.639 | 2.642 | 1.583 | 92.98 | 33.54 |
| 650.0 | 25.017 | 3703 | 3302 | 6.784 | 2.581 | 1.406 | 98.94 | 35.54 |
| 700.0 | 26.747 | 3831 | 3403 | 6.919 | 2.548 | 1.273 | 105.1 | 37.50 |
| 750.0 | 28.426 | 3958 | 3503 | 7.047 | 2.534 | 1.167 | 111.4 | 39.41 |
| 800.0 | 30.067 | 4084 | 3603 | 7.168 | 2.531 | 1.081 | 117.7 | 41.28 |
| 850.0 | 31.679 | 4211 | 3704 | 7.283 | 2.536 | 1.009 | 124.0 | 43.11 |
| 900.0 | 33.266 | 4338 | 3806 | 7.394 | 2.547 | 0.948 | 130.3 | 44.90 |
| 950.0 | 34.835 | 4466 | 3908 | 7.500 | 2.562 | 0.896 | 136.5 | 46.66 |
| 1000.0 | 36.388 | 4594 | 4012 | 7.603 | 2.579 | 0.850 | 142.7 | 48.38 |

Table 1 (cont.) $p = 17.0\,\text{MPa} = 170\ \text{bar}$

| $t$ | $v$ | $h$ | $u$ | $s$ | $c_p$ | $\alpha_p$ | $\lambda$ | $\eta$ |
|---|---|---|---|---|---|---|---|---|
| °C | dm³/kg | kJ/kg | kJ/kg | kJ/kg K | kJ/kg K | $10^{-3}$/K | mW/K m | $\mu$Pa s |
| 0.0 | 0.9918 | 17.1 | 0.2 | 0.0006 | 4.143 | −0.014 | 570.4 | 1753 |
| 5.0 | 0.9919 | 37.8 | 20.9 | 0.076 | 4.128 | 0.063 | 579.2 | 1493 |
| 10.0 | 0.9924 | 58.4 | 41.5 | 0.149 | 4.126 | 0.127 | 588.3 | 1289 |
| 15.0 | 0.9932 | 79.0 | 62.2 | 0.221 | 4.129 | 0.183 | 597.3 | 1127 |
| 20.0 | 0.9942 | 99.7 | 82.8 | 0.292 | 4.133 | 0.233 | 606.2 | 995.7 |
| 25.0 | 0.9955 | 120.4 | 103.5 | 0.362 | 4.137 | 0.277 | 614.8 | 887.2 |
| 30.0 | 0.9970 | 141.1 | 124.1 | 0.431 | 4.140 | 0.318 | 623.1 | 796.6 |
| 35.0 | 0.9986 | 161.8 | 144.8 | 0.499 | 4.142 | 0.355 | 630.9 | 720.1 |
| 40.0 | 1.0005 | 182.5 | 165.5 | 0.566 | 4.143 | 0.390 | 638.3 | 654.9 |
| 45.0 | 1.0025 | 203.2 | 186.2 | 0.631 | 4.144 | 0.424 | 645.1 | 598.8 |
| 50.0 | 1.0048 | 223.9 | 206.9 | 0.696 | 4.145 | 0.455 | 651.4 | 550.2 |
| 60.0 | 1.0096 | 265.4 | 248.2 | 0.822 | 4.147 | 0.514 | 662.4 | 470.5 |
| 70.0 | 1.0151 | 306.9 | 289.6 | 0.945 | 4.152 | 0.570 | 671.4 | 408.4 |
| 80.0 | 1.0212 | 348.4 | 331.1 | 1.064 | 4.159 | 0.623 | 678.6 | 359.0 |
| 90.0 | 1.0278 | 390.1 | 372.6 | 1.181 | 4.168 | 0.673 | 684.2 | 319.1 |
| 100.0 | 1.0350 | 431.8 | 414.2 | 1.294 | 4.180 | 0.723 | 688.3 | 286.4 |
| 125.0 | 1.0555 | 536.7 | 518.8 | 1.566 | 4.216 | 0.845 | 693.7 | 226.5 |
| 150.0 | 1.0797 | 642.7 | 624.3 | 1.824 | 4.262 | 0.971 | 693.1 | 186.7 |
| 175.0 | 1.1081 | 749.9 | 731.1 | 2.070 | 4.323 | 1.110 | 687.5 | 158.7 |
| 200.0 | 1.1416 | 859.0 | 839.6 | 2.307 | 4.408 | 1.272 | 676.6 | 138.2 |
| 225.0 | 1.1813 | 970.6 | 950.5 | 2.537 | 4.528 | 1.475 | 660.2 | 122.4 |
| 250.0 | 1.2296 | 1085 | 1064 | 2.763 | 4.704 | 1.744 | 637.3 | 109.7 |
| 275.0 | 1.2902 | 1206 | 1184 | 2.988 | 4.968 | 2.131 | 606.4 | 98.92 |
| 300.0 | 1.3705 | 1335 | 1312 | 3.218 | 5.402 | 2.762 | 565.8 | 89.11 |
| 325.0 | 1.4891 | 1480 | 1454 | 3.465 | 6.272 | 4.067 | 514.4 | 79.19 |
| 350.0 | 1.7269 | 1666 | 1636 | 3.769 | 9.702 | 9.645 | 450.0 | 66.45 |
| 352.34 | saturation | | | | | | | |
| liquid | 1.7698 | 1689 | 1659 | 3.807 | 10.82 | 11.54 | 442.7 | 64.71 |
| vapour | 8.3733 | 2547 | 2404 | 5.178 | 17.62 | 24.51 | 142.4 | 24.18 |
| 375.0 | 11.168 | 2775 | 2585 | 5.537 | 6.879 | 7.884 | 99.10 | 24.34 |
| 400.0 | 13.025 | 2917 | 2695 | 5.752 | 4.846 | 4.948 | 88.44 | 25.26 |
| 450.0 | 15.759 | 3119 | 2851 | 6.043 | 3.518 | 3.041 | 83.38 | 27.36 |
| 500.0 | 17.967 | 3281 | 2976 | 6.260 | 3.033 | 2.291 | 84.82 | 29.49 |
| 550.0 | 19.925 | 3427 | 3088 | 6.442 | 2.800 | 1.880 | 88.81 | 31.58 |
| 600.0 | 21.736 | 3563 | 3194 | 6.603 | 2.675 | 1.616 | 94.07 | 33.63 |
| 650.0 | 23.450 | 3695 | 3296 | 6.750 | 2.606 | 1.430 | 99.95 | 35.63 |
| 700.0 | 25.095 | 3824 | 3398 | 6.886 | 2.568 | 1.290 | 106.1 | 37.58 |
| 750.0 | 26.691 | 3952 | 3498 | 7.014 | 2.549 | 1.180 | 112.3 | 39.49 |
| 800.0 | 28.248 | 4079 | 3599 | 7.136 | 2.544 | 1.091 | 118.6 | 41.35 |
| 850.0 | 29.775 | 4207 | 3700 | 7.252 | 2.547 | 1.017 | 124.8 | 43.18 |
| 900.0 | 31.278 | 4334 | 3802 | 7.363 | 2.556 | 0.955 | 131.0 | 44.97 |
| 950.0 | 32.762 | 4462 | 3905 | 7.470 | 2.570 | 0.901 | 137.1 | 46.72 |
| 1000.0 | 34.230 | 4591 | 4009 | 7.573 | 2.586 | 0.854 | 143.2 | 48.44 |

Table 1 (cont.)

$$p = 18.0\,\text{MPa} = 180\ \text{bar}$$

| $t$ | $v$ | $h$ | $u$ | $s$ | $c_p$ | $\alpha_p$ | $\lambda$ | $\eta$ |
|---|---|---|---|---|---|---|---|---|
| °C | dm³/kg | kJ/kg | kJ/kg | kJ/kg K | kJ/kg K | $10^{-3}$/K | mW/K m | $\mu$Pa s |
| 0.0 | 0.9913 | 18.1 | 0.3 | 0.0006 | 4.139 | −0.010 | 570.9 | 1751 |
| 5.0 | 0.9915 | 38.8 | 20.9 | 0.076 | 4.124 | 0.065 | 579.7 | 1491 |
| 10.0 | 0.9920 | 59.4 | 41.5 | 0.149 | 4.123 | 0.129 | 588.7 | 1288 |
| 15.0 | 0.9927 | 80.0 | 62.1 | 0.221 | 4.126 | 0.185 | 597.8 | 1126 |
| 20.0 | 0.9938 | 100.6 | 82.7 | 0.292 | 4.131 | 0.234 | 606.7 | 995.4 |
| 25.0 | 0.9951 | 121.3 | 103.4 | 0.362 | 4.135 | 0.278 | 615.3 | 887.1 |
| 30.0 | 0.9965 | 142.0 | 124.0 | 0.431 | 4.138 | 0.318 | 623.6 | 796.6 |
| 35.0 | 0.9982 | 162.7 | 144.7 | 0.499 | 4.140 | 0.356 | 631.4 | 720.2 |
| 40.0 | 1.0001 | 183.4 | 165.4 | 0.565 | 4.141 | 0.391 | 638.7 | 655.0 |
| 45.0 | 1.0021 | 204.1 | 186.0 | 0.631 | 4.142 | 0.424 | 645.6 | 599.0 |
| 50.0 | 1.0043 | 224.8 | 206.7 | 0.695 | 4.143 | 0.455 | 651.9 | 550.4 |
| 60.0 | 1.0092 | 266.2 | 248.1 | 0.822 | 4.145 | 0.514 | 662.9 | 470.7 |
| 70.0 | 1.0147 | 307.7 | 289.4 | 0.944 | 4.150 | 0.569 | 671.9 | 408.6 |
| 80.0 | 1.0208 | 349.2 | 330.9 | 1.064 | 4.157 | 0.622 | 679.1 | 359.3 |
| 90.0 | 1.0274 | 390.8 | 372.3 | 1.180 | 4.166 | 0.672 | 684.7 | 319.4 |
| 100.0 | 1.0346 | 432.6 | 413.9 | 1.293 | 4.178 | 0.721 | 688.9 | 286.7 |
| 125.0 | 1.0550 | 537.4 | 518.4 | 1.565 | 4.214 | 0.843 | 694.3 | 226.8 |
| 150.0 | 1.0791 | 643.3 | 623.9 | 1.823 | 4.259 | 0.968 | 693.8 | 186.9 |
| 175.0 | 1.1074 | 750.5 | 730.6 | 2.069 | 4.319 | 1.105 | 688.2 | 159.0 |
| 200.0 | 1.1407 | 859.5 | 838.9 | 2.306 | 4.403 | 1.266 | 677.5 | 138.4 |
| 225.0 | 1.1802 | 970.9 | 949.7 | 2.535 | 4.522 | 1.466 | 661.2 | 122.6 |
| 250.0 | 1.2281 | 1086 | 1063 | 2.761 | 4.694 | 1.730 | 638.5 | 110.0 |
| 275.0 | 1.2880 | 1206 | 1183 | 2.985 | 4.952 | 2.108 | 607.9 | 99.20 |
| 300.0 | 1.3671 | 1334 | 1310 | 3.215 | 5.370 | 2.716 | 567.8 | 89.45 |
| 325.0 | 1.4825 | 1478 | 1451 | 3.459 | 6.188 | 3.938 | 517.2 | 79.66 |
| 350.0 | 1.7028 | 1658 | 1627 | 3.754 | 9.025 | 8.502 | 454.7 | 67.52 |
| 357.04 | saturation | | | | | | | |
| liquid | 1.8399 | 1732 | 1698 | 3.871 | 12.78 | 14.94 | 433.0 | 62.16 |
| vapour | 7.5046 | 2509 | 2374 | 5.105 | 21.83 | 31.25 | 162.0 | 25.05 |
| 375.0 | 9.9590 | 2726 | 2546 | 5.445 | 8.095 | 9.645 | 108.9 | 24.70 |
| 400.0 | 11.905 | 2885 | 2671 | 5.687 | 5.267 | 5.512 | 93.23 | 25.47 |
| 450.0 | 14.631 | 3100 | 2837 | 5.995 | 3.662 | 3.217 | 85.73 | 27.50 |
| 500.0 | 16.784 | 3267 | 2965 | 6.219 | 3.108 | 2.378 | 86.42 | 29.60 |
| 550.0 | 18.675 | 3416 | 3080 | 6.405 | 2.848 | 1.932 | 90.10 | 31.68 |
| 600.0 | 20.415 | 3554 | 3187 | 6.569 | 2.709 | 1.650 | 95.20 | 33.72 |
| 650.0 | 22.056 | 3688 | 3291 | 6.717 | 2.631 | 1.453 | 101.0 | 35.72 |
| 700.0 | 23.628 | 3818 | 3393 | 6.855 | 2.587 | 1.307 | 107.1 | 37.66 |
| 750.0 | 25.149 | 3947 | 3494 | 6.984 | 2.565 | 1.193 | 113.3 | 39.57 |
| 800.0 | 26.631 | 4075 | 3595 | 7.106 | 2.557 | 1.101 | 119.4 | 41.43 |
| 850.0 | 28.083 | 4202 | 3697 | 7.222 | 2.558 | 1.025 | 125.6 | 43.25 |
| 900.0 | 29.511 | 4330 | 3799 | 7.334 | 2.565 | 0.961 | 131.7 | 45.04 |
| 950.0 | 30.920 | 4459 | 3902 | 7.441 | 2.577 | 0.906 | 137.8 | 46.79 |
| 1000.0 | 32.312 | 4588 | 4007 | 7.545 | 2.593 | 0.858 | 143.8 | 48.50 |

Table 1 (cont.)

$$p = 19.0 \, \text{MPa} = 190 \, \text{bar}$$

| $t$ | $v$ | $h$ | $u$ | $s$ | $c_p$ | $\alpha_p$ | $\lambda$ | $\eta$ |
|---|---|---|---|---|---|---|---|---|
| °C | dm³/kg | kJ/kg | kJ/kg | kJ/kg K | kJ/kg K | $10^{-3}$/K | mW/K m | $\mu$Pa s |
| 0.0 | 0.9908 | 19.1 | 0.3 | 0.0007 | 4.135 | −0.006 | 571.5 | 1749 |
| 5.0 | 0.9910 | 39.7 | 20.9 | 0.075 | 4.121 | 0.068 | 580.2 | 1490 |
| 10.0 | 0.9915 | 60.3 | 41.5 | 0.149 | 4.119 | 0.132 | 589.2 | 1287 |
| 15.0 | 0.9923 | 80.9 | 62.1 | 0.221 | 4.123 | 0.187 | 598.2 | 1126 |
| 20.0 | 0.9933 | 101.6 | 82.7 | 0.292 | 4.128 | 0.235 | 607.1 | 995.1 |
| 25.0 | 0.9946 | 122.2 | 103.3 | 0.362 | 4.132 | 0.279 | 615.8 | 886.9 |
| 30.0 | 0.9961 | 142.9 | 123.9 | 0.431 | 4.136 | 0.319 | 624.0 | 796.6 |
| 35.0 | 0.9978 | 163.6 | 144.6 | 0.498 | 4.138 | 0.356 | 631.9 | 720.2 |
| 40.0 | 0.9997 | 184.2 | 165.3 | 0.565 | 4.139 | 0.391 | 639.2 | 655.2 |
| 45.0 | 1.0017 | 204.9 | 185.9 | 0.630 | 4.140 | 0.424 | 646.0 | 599.2 |
| 50.0 | 1.0039 | 225.6 | 206.6 | 0.695 | 4.141 | 0.455 | 652.3 | 550.6 |
| 60.0 | 1.0088 | 267.1 | 247.9 | 0.821 | 4.143 | 0.514 | 663.4 | 471.0 |
| 70.0 | 1.0143 | 308.5 | 289.2 | 0.944 | 4.148 | 0.568 | 672.4 | 408.9 |
| 80.0 | 1.0203 | 350.0 | 330.6 | 1.063 | 4.155 | 0.621 | 679.6 | 359.5 |
| 90.0 | 1.0269 | 391.6 | 372.1 | 1.179 | 4.164 | 0.671 | 685.2 | 319.6 |
| 100.0 | 1.0341 | 433.3 | 413.7 | 1.292 | 4.176 | 0.720 | 689.4 | 286.9 |
| 125.0 | 1.0545 | 538.1 | 518.1 | 1.564 | 4.211 | 0.840 | 694.9 | 227.1 |
| 150.0 | 1.0785 | 644.0 | 623.5 | 1.822 | 4.256 | 0.965 | 694.5 | 187.2 |
| 175.0 | 1.1067 | 751.1 | 730.0 | 2.068 | 4.316 | 1.101 | 688.9 | 159.2 |
| 200.0 | 1.1398 | 859.9 | 838.3 | 2.304 | 4.398 | 1.260 | 678.3 | 138.6 |
| 225.0 | 1.1790 | 971.3 | 948.9 | 2.534 | 4.515 | 1.458 | 662.1 | 122.9 |
| 250.0 | 1.2266 | 1086 | 1062 | 2.759 | 4.684 | 1.717 | 639.6 | 110.2 |
| 275.0 | 1.2859 | 1206 | 1181 | 2.983 | 4.936 | 2.085 | 609.3 | 99.49 |
| 300.0 | 1.3638 | 1334 | 1308 | 3.211 | 5.340 | 2.672 | 569.7 | 89.79 |
| 325.0 | 1.4762 | 1476 | 1448 | 3.453 | 6.111 | 3.820 | 519.9 | 80.13 |
| 350.0 | 1.6824 | 1651 | 1619 | 3.740 | 8.526 | 7.668 | 459.1 | 68.47 |
| 361.52 | | | | saturation | | | | |
| liquid | 1.9252 | 1776 | 1740 | 3.939 | 16.02 | 20.66 | 424.5 | 59.42 |
| vapour | 6.6815 | 2466 | 2339 | 5.026 | 28.50 | 42.15 | 187.9 | 26.10 |
| 375.0 | 8.8066 | 2669 | 2502 | 5.343 | 9.908 | 12.34 | 122.3 | 25.20 |
| 400.0 | 10.885 | 2852 | 2645 | 5.621 | 5.767 | 6.193 | 98.82 | 25.73 |
| 450.0 | 13.618 | 3081 | 2822 | 5.949 | 3.816 | 3.407 | 88.28 | 27.65 |
| 500.0 | 15.724 | 3253 | 2955 | 6.180 | 3.187 | 2.468 | 88.11 | 29.72 |
| 550.0 | 17.556 | 3405 | 3071 | 6.370 | 2.897 | 1.985 | 91.43 | 31.79 |
| 600.0 | 19.233 | 3545 | 3180 | 6.536 | 2.743 | 1.684 | 96.36 | 33.82 |
| 650.0 | 20.809 | 3680 | 3285 | 6.686 | 2.656 | 1.477 | 102.1 | 35.81 |
| 700.0 | 22.315 | 3811 | 3387 | 6.824 | 2.607 | 1.324 | 108.1 | 37.75 |
| 750.0 | 23.769 | 3941 | 3489 | 6.954 | 2.581 | 1.206 | 114.2 | 39.65 |
| 800.0 | 25.185 | 4070 | 3591 | 7.077 | 2.569 | 1.111 | 120.3 | 41.51 |
| 850.0 | 26.570 | 4198 | 3693 | 7.194 | 2.568 | 1.033 | 126.4 | 43.32 |
| 900.0 | 27.930 | 4327 | 3796 | 7.306 | 2.574 | 0.967 | 132.5 | 45.10 |
| 950.0 | 29.271 | 4456 | 3900 | 7.414 | 2.585 | 0.911 | 138.5 | 46.85 |
| 1000.0 | 30.597 | 4585 | 4004 | 7.518 | 2.599 | 0.862 | 144.4 | 48.56 |

Table 1 (cont.)

$p = 20.0\,\text{MPa} = 200\ \text{bar}$

| $t$ | $v$ | $h$ | $u$ | $s$ | $c_\mathrm{p}$ | $\alpha_\mathrm{p}$ | $\lambda$ | $\eta$ |
|---|---|---|---|---|---|---|---|---|
| °C | dm³/kg | kJ/kg | kJ/kg | kJ/kg K | kJ/kg K | $10^{-3}$/K | mW/K m | $\mu$Pa s |
| 0.0 | 0.9904 | 20.1 | 0.3 | 0.0007 | 4.130 | −0.003 | 572.0 | 1747 |
| 5.0 | 0.9905 | 40.7 | 20.9 | 0.075 | 4.117 | 0.071 | 580.8 | 1489 |
| 10.0 | 0.9911 | 61.3 | 41.5 | 0.149 | 4.116 | 0.134 | 589.7 | 1287 |
| 15.0 | 0.9919 | 81.9 | 62.0 | 0.221 | 4.120 | 0.188 | 598.7 | 1125 |
| 20.0 | 0.9929 | 102.5 | 82.6 | 0.292 | 4.125 | 0.237 | 607.6 | 994.8 |
| 25.0 | 0.9942 | 123.1 | 103.2 | 0.362 | 4.130 | 0.280 | 616.2 | 886.8 |
| 30.0 | 0.9957 | 143.8 | 123.9 | 0.430 | 4.133 | 0.320 | 624.5 | 796.5 |
| 35.0 | 0.9974 | 164.4 | 144.5 | 0.498 | 4.135 | 0.357 | 632.3 | 720.3 |
| 40.0 | 0.9992 | 185.1 | 165.1 | 0.564 | 4.137 | 0.391 | 639.7 | 655.3 |
| 45.0 | 1.0013 | 205.8 | 185.8 | 0.630 | 4.138 | 0.424 | 646.5 | 599.3 |
| 50.0 | 1.0035 | 226.5 | 206.4 | 0.695 | 4.139 | 0.455 | 652.8 | 550.8 |
| 60.0 | 1.0084 | 267.9 | 247.7 | 0.821 | 4.141 | 0.513 | 663.8 | 471.2 |
| 70.0 | 1.0138 | 309.3 | 289.1 | 0.943 | 4.146 | 0.568 | 672.9 | 409.1 |
| 80.0 | 1.0199 | 350.8 | 330.4 | 1.062 | 4.153 | 0.620 | 680.1 | 359.8 |
| 90.0 | 1.0265 | 392.4 | 371.9 | 1.179 | 4.162 | 0.670 | 685.7 | 319.9 |
| 100.0 | 1.0336 | 434.1 | 413.4 | 1.292 | 4.174 | 0.718 | 690.0 | 287.2 |
| 125.0 | 1.0539 | 538.8 | 517.8 | 1.563 | 4.209 | 0.838 | 695.5 | 227.3 |
| 150.0 | 1.0779 | 644.6 | 623.0 | 1.821 | 4.254 | 0.962 | 695.1 | 187.4 |
| 175.0 | 1.1059 | 751.6 | 729.5 | 2.067 | 4.312 | 1.097 | 689.7 | 159.4 |
| 200.0 | 1.1389 | 860.4 | 837.6 | 2.303 | 4.394 | 1.255 | 679.1 | 138.9 |
| 225.0 | 1.1779 | 971.6 | 948.0 | 2.532 | 4.509 | 1.449 | 663.1 | 123.1 |
| 250.0 | 1.2251 | 1086 | 1061 | 2.757 | 4.674 | 1.704 | 640.8 | 110.5 |
| 275.0 | 1.2838 | 1205 | 1180 | 2.980 | 4.920 | 2.064 | 610.8 | 99.77 |
| 300.0 | 1.3605 | 1333 | 1306 | 3.207 | 5.311 | 2.630 | 571.6 | 90.12 |
| 325.0 | 1.4702 | 1474 | 1444 | 3.448 | 6.039 | 3.711 | 522.5 | 80.58 |
| 350.0 | 1.6645 | 1645 | 1612 | 3.728 | 8.138 | 7.028 | 463.3 | 69.33 |
| 365.80 | | | | saturation | | | | |
| liquid | 2.0360 | 1826 | 1786 | 4.015 | 22.37 | 32.07 | 419.4 | 56.30 |
| vapour | 5.8738 | 2413 | 2296 | 4.933 | 40.99 | 63.01 | 225.5 | 27.47 |
| 375.0 | 7.6675 | 2600 | 2447 | 5.225 | 12.98 | 17.05 | 142.1 | 25.92 |
| 400.0 | 9.9458 | 2816 | 2617 | 5.552 | 6.371 | 7.030 | 105.4 | 26.03 |
| 450.0 | 12.701 | 3060 | 2806 | 5.903 | 3.982 | 3.613 | 91.03 | 27.81 |
| 500.0 | 14.769 | 3239 | 2944 | 6.142 | 3.269 | 2.563 | 89.89 | 29.85 |
| 550.0 | 16.549 | 3393 | 3062 | 6.335 | 2.948 | 2.039 | 92.81 | 31.90 |
| 600.0 | 18.169 | 3536 | 3173 | 6.504 | 2.778 | 1.720 | 97.57 | 33.92 |
| 650.0 | 19.687 | 3672 | 3279 | 6.656 | 2.682 | 1.502 | 103.2 | 35.90 |
| 700.0 | 21.133 | 3805 | 3382 | 6.796 | 2.627 | 1.342 | 109.1 | 37.84 |
| 750.0 | 22.528 | 3936 | 3485 | 6.926 | 2.597 | 1.219 | 115.2 | 39.73 |
| 800.0 | 23.883 | 4065 | 3587 | 7.050 | 2.583 | 1.121 | 121.2 | 41.58 |
| 850.0 | 25.207 | 4194 | 3690 | 7.167 | 2.579 | 1.041 | 127.2 | 43.40 |
| 900.0 | 26.508 | 4323 | 3793 | 7.280 | 2.584 | 0.973 | 133.2 | 45.17 |
| 950.0 | 27.788 | 4452 | 3897 | 7.388 | 2.593 | 0.916 | 139.1 | 46.92 |
| 1000.0 | 29.053 | 4582 | 4001 | 7.492 | 2.606 | 0.866 | 145.0 | 48.62 |

Table 1 (cont.)

$p = 21.0\,\mathrm{MPa} = 210\ \mathrm{bar}$

| $t$ | $v$ | $h$ | $u$ | $s$ | $c_\mathrm{p}$ | $\alpha_\mathrm{p}$ | $\lambda$ | $\eta$ |
|---|---|---|---|---|---|---|---|---|
| °C | dm³/kg | kJ/kg | kJ/kg | kJ/kg K | kJ/kg K | $10^{-3}$/K | mW/K m | $\mu$Pa s |
| 0.0 | 0.9899 | 21.1 | 0.3 | 0.0007 | 4.126 | 0.001 | 572.6 | 1745 |
| 5.0 | 0.9901 | 41.7 | 20.9 | 0.075 | 4.113 | 0.074 | 581.3 | 1487 |
| 10.0 | 0.9906 | 62.2 | 41.4 | 0.149 | 4.113 | 0.136 | 590.2 | 1286 |
| 15.0 | 0.9914 | 82.8 | 62.0 | 0.221 | 4.117 | 0.190 | 599.2 | 1125 |
| 20.0 | 0.9925 | 103.4 | 82.6 | 0.292 | 4.123 | 0.238 | 608.0 | 994.5 |
| 25.0 | 0.9938 | 124.0 | 103.2 | 0.361 | 4.127 | 0.281 | 616.7 | 886.7 |
| 30.0 | 0.9953 | 144.7 | 123.8 | 0.430 | 4.131 | 0.321 | 624.9 | 796.5 |
| 35.0 | 0.9970 | 165.3 | 144.4 | 0.498 | 4.133 | 0.357 | 632.8 | 720.4 |
| 40.0 | 0.9988 | 186.0 | 165.0 | 0.564 | 4.135 | 0.392 | 640.1 | 655.4 |
| 45.0 | 1.0009 | 206.7 | 185.7 | 0.630 | 4.136 | 0.424 | 647.0 | 599.5 |
| 50.0 | 1.0031 | 227.4 | 206.3 | 0.694 | 4.137 | 0.455 | 653.3 | 551.0 |
| 60.0 | 1.0079 | 268.7 | 247.6 | 0.820 | 4.139 | 0.513 | 664.3 | 471.5 |
| 70.0 | 1.0134 | 310.1 | 288.9 | 0.943 | 4.144 | 0.567 | 673.4 | 409.4 |
| 80.0 | 1.0194 | 351.6 | 330.2 | 1.062 | 4.151 | 0.619 | 680.6 | 360.1 |
| 90.0 | 1.0260 | 393.2 | 371.6 | 1.178 | 4.160 | 0.669 | 686.3 | 320.2 |
| 100.0 | 1.0331 | 434.8 | 413.1 | 1.291 | 4.172 | 0.717 | 690.5 | 287.5 |
| 125.0 | 1.0534 | 539.5 | 517.4 | 1.563 | 4.207 | 0.836 | 696.1 | 227.6 |
| 150.0 | 1.0773 | 645.2 | 622.6 | 1.820 | 4.251 | 0.959 | 695.8 | 187.7 |
| 175.0 | 1.1052 | 752.2 | 729.0 | 2.066 | 4.309 | 1.093 | 690.4 | 159.7 |
| 200.0 | 1.1380 | 860.9 | 837.0 | 2.302 | 4.389 | 1.249 | 679.9 | 139.1 |
| 225.0 | 1.1768 | 971.9 | 947.2 | 2.530 | 4.503 | 1.441 | 664.0 | 123.4 |
| 250.0 | 1.2236 | 1086 | 1060 | 2.754 | 4.665 | 1.691 | 641.9 | 110.7 |
| 275.0 | 1.2817 | 1205 | 1178 | 2.977 | 4.905 | 2.043 | 612.2 | 100.0 |
| 300.0 | 1.3573 | 1332 | 1304 | 3.204 | 5.283 | 2.590 | 573.5 | 90.45 |
| 325.0 | 1.4644 | 1472 | 1441 | 3.442 | 5.973 | 3.612 | 525.1 | 81.01 |
| 350.0 | 1.6486 | 1640 | 1605 | 3.716 | 7.825 | 6.517 | 467.2 | 70.12 |
| 369.88 | | | | saturation | | | | |
| liquid | 2.2003 | 1887 | 1841 | 4.106 | 40.82 | 66.07 | 427.2 | 52.39 |
| vapour | 5.0204 | 2342 | 2237 | 4.814 | 76.58 | 123.1 | 296.0 | 29.49 |
| 375.0 | 6.4607 | 2510 | 2374 | 5.073 | 19.70 | 27.76 | 176.2 | 27.13 |
| 400.0 | 9.0738 | 2778 | 2587 | 5.481 | 7.114 | 8.079 | 113.3 | 26.41 |
| 450.0 | 11.868 | 3040 | 2790 | 5.857 | 4.162 | 3.838 | 94.02 | 28.00 |
| 500.0 | 13.903 | 3224 | 2932 | 6.104 | 3.355 | 2.662 | 91.76 | 29.99 |
| 550.0 | 15.637 | 3382 | 3054 | 6.302 | 3.000 | 2.095 | 94.25 | 32.02 |
| 600.0 | 17.206 | 3527 | 3166 | 6.473 | 2.813 | 1.755 | 98.80 | 34.03 |
| 650.0 | 18.671 | 3665 | 3273 | 6.627 | 2.708 | 1.526 | 104.3 | 36.00 |
| 700.0 | 20.064 | 3799 | 3377 | 6.768 | 2.647 | 1.360 | 110.2 | 37.93 |
| 750.0 | 21.404 | 3930 | 3480 | 6.899 | 2.613 | 1.232 | 116.1 | 39.82 |
| 800.0 | 22.705 | 4060 | 3583 | 7.024 | 2.596 | 1.131 | 122.1 | 41.67 |
| 850.0 | 23.975 | 4190 | 3686 | 7.142 | 2.590 | 1.048 | 128.1 | 43.47 |
| 900.0 | 25.221 | 4319 | 3790 | 7.254 | 2.593 | 0.979 | 134.0 | 45.25 |
| 950.0 | 26.446 | 4449 | 3894 | 7.363 | 2.601 | 0.921 | 139.8 | 46.98 |
| 1000.0 | 27.656 | 4579 | 3999 | 7.467 | 2.613 | 0.870 | 145.6 | 48.68 |

Table 1 (cont.)

**p = 22.0 MPa = 220 bar**

| $t$ | $v$ | $h$ | $u$ | $s$ | $c_p$ | $\alpha_p$ | $\lambda$ | $\eta$ |
|---|---|---|---|---|---|---|---|---|
| °C | dm³/kg | kJ/kg | kJ/kg | kJ/kg K | kJ/kg K | $10^{-3}$/K | mW/K m | $\mu$Pa s |
| 0.0 | 0.9894 | 22.1 | 0.3 | 0.0007 | 4.122 | 0.004 | 573.1 | 1743 |
| 5.0 | 0.9896 | 42.6 | 20.9 | 0.075 | 4.109 | 0.077 | 581.8 | 1486 |
| 10.0 | 0.9902 | 63.2 | 41.4 | 0.148 | 4.109 | 0.138 | 590.7 | 1285 |
| 15.0 | 0.9910 | 83.7 | 61.9 | 0.220 | 4.114 | 0.192 | 599.7 | 1124 |
| 20.0 | 0.9920 | 104.3 | 82.5 | 0.291 | 4.120 | 0.239 | 608.5 | 994.3 |
| 25.0 | 0.9933 | 124.9 | 103.1 | 0.361 | 4.125 | 0.282 | 617.1 | 886.5 |
| 30.0 | 0.9948 | 145.6 | 123.7 | 0.430 | 4.129 | 0.321 | 625.4 | 796.5 |
| 35.0 | 0.9965 | 166.2 | 144.3 | 0.497 | 4.131 | 0.358 | 633.2 | 720.4 |
| 40.0 | 0.9984 | 186.9 | 164.9 | 0.564 | 4.133 | 0.392 | 640.6 | 655.5 |
| 45.0 | 1.0004 | 207.5 | 185.5 | 0.629 | 4.134 | 0.424 | 647.4 | 599.7 |
| 50.0 | 1.0026 | 228.2 | 206.2 | 0.694 | 4.135 | 0.455 | 653.7 | 551.2 |
| 60.0 | 1.0075 | 269.6 | 247.4 | 0.820 | 4.137 | 0.512 | 664.8 | 471.7 |
| 70.0 | 1.0130 | 311.0 | 288.7 | 0.942 | 4.142 | 0.566 | 673.9 | 409.7 |
| 80.0 | 1.0190 | 352.4 | 330.0 | 1.061 | 4.149 | 0.618 | 681.1 | 360.3 |
| 90.0 | 1.0256 | 393.9 | 371.4 | 1.177 | 4.158 | 0.667 | 686.8 | 320.4 |
| 100.0 | 1.0327 | 435.6 | 412.9 | 1.290 | 4.169 | 0.716 | 691.0 | 287.7 |
| 125.0 | 1.0529 | 540.2 | 517.1 | 1.562 | 4.204 | 0.834 | 696.7 | 227.8 |
| 150.0 | 1.0767 | 645.9 | 622.2 | 1.819 | 4.248 | 0.956 | 696.4 | 187.9 |
| 175.0 | 1.1045 | 752.8 | 728.5 | 2.064 | 4.306 | 1.089 | 691.1 | 159.9 |
| 200.0 | 1.1371 | 861.3 | 836.3 | 2.300 | 4.385 | 1.243 | 680.8 | 139.3 |
| 225.0 | 1.1757 | 972.3 | 946.4 | 2.529 | 4.496 | 1.433 | 665.0 | 123.6 |
| 250.0 | 1.2221 | 1086 | 1059 | 2.752 | 4.656 | 1.679 | 643.0 | 111.0 |
| 275.0 | 1.2797 | 1205 | 1177 | 2.975 | 4.890 | 2.022 | 613.6 | 100.3 |
| 300.0 | 1.3542 | 1332 | 1302 | 3.200 | 5.256 | 2.552 | 575.3 | 90.77 |
| 325.0 | 1.4589 | 1470 | 1438 | 3.437 | 5.912 | 3.519 | 527.6 | 81.44 |
| 350.0 | 1.6343 | 1635 | 1599 | 3.706 | 7.567 | 6.097 | 471.0 | 70.86 |
| 373.77 | | | | saturation | | | | |
| liquid | 2.7011 | 2012 | 1953 | 4.296 | 1403 | 2801 | 907.1 | 47.04 |
| vapour | 3.6526 | 2176 | 2095 | 4.549 | 4041 | 6865 | 1408 | 38.18 |
| 375.0 | 4.8643 | 2349 | 2242 | 4.817 | 55.30 | 87.60 | 275.7 | 30.21 |
| 400.0 | 8.2555 | 2736 | 2554 | 5.405 | 8.054 | 9.429 | 122.9 | 26.87 |
| 450.0 | 11.107 | 3018 | 2774 | 5.811 | 4.356 | 4.082 | 97.26 | 28.20 |
| 500.0 | 13.115 | 3210 | 2921 | 6.068 | 3.445 | 2.766 | 93.72 | 30.14 |
| 550.0 | 14.808 | 3371 | 3045 | 6.270 | 3.053 | 2.153 | 95.75 | 32.14 |
| 600.0 | 16.331 | 3518 | 3159 | 6.443 | 2.849 | 1.792 | 100.1 | 34.14 |
| 650.0 | 17.748 | 3657 | 3267 | 6.599 | 2.734 | 1.551 | 105.4 | 36.10 |
| 700.0 | 19.092 | 3792 | 3372 | 6.741 | 2.667 | 1.378 | 111.2 | 38.02 |
| 750.0 | 20.384 | 3924 | 3476 | 6.874 | 2.629 | 1.246 | 117.1 | 39.90 |
| 800.0 | 21.635 | 4055 | 3579 | 6.998 | 2.609 | 1.141 | 123.1 | 41.75 |
| 850.0 | 22.855 | 4185 | 3683 | 7.117 | 2.601 | 1.056 | 129.0 | 43.55 |
| 900.0 | 24.051 | 4315 | 3786 | 7.230 | 2.602 | 0.986 | 134.8 | 45.32 |
| 950.0 | 25.227 | 4446 | 3891 | 7.339 | 2.609 | 0.925 | 140.6 | 47.05 |
| 1000.0 | 26.387 | 4576 | 3996 | 7.444 | 2.620 | 0.874 | 146.3 | 48.75 |

Table 1 (cont.)

$$p = 23.0\,\text{MPa} = 230\ \text{bar}$$

| $t$ | $v$ | $h$ | $u$ | $s$ | $c_p$ | $\alpha_p$ | $\lambda$ | $\eta$ |
|---|---|---|---|---|---|---|---|---|
| °C | dm³/kg | kJ/kg | kJ/kg | kJ/kg K | kJ/kg K | 10⁻³/K | mW/K m | µPa s |
| 0.0 | 0.9889 | 23.1 | 0.3 | 0.0007 | 4.117 | 0.008 | 573.7 | 1741 |
| 5.0 | 0.9892 | 43.6 | 20.9 | 0.075 | 4.106 | 0.079 | 582.3 | 1485 |
| 10.0 | 0.9897 | 64.1 | 41.4 | 0.148 | 4.106 | 0.140 | 591.2 | 1284 |
| 15.0 | 0.9905 | 84.7 | 61.9 | 0.220 | 4.111 | 0.194 | 600.1 | 1124 |
| 20.0 | 0.9916 | 105.2 | 82.4 | 0.291 | 4.117 | 0.241 | 609.0 | 994.0 |
| 25.0 | 0.9929 | 125.8 | 103.0 | 0.361 | 4.122 | 0.283 | 617.6 | 886.4 |
| 30.0 | 0.9944 | 146.5 | 123.6 | 0.429 | 4.126 | 0.322 | 625.8 | 796.5 |
| 35.0 | 0.9961 | 167.1 | 144.2 | 0.497 | 4.129 | 0.358 | 633.7 | 720.5 |
| 40.0 | 0.9980 | 187.8 | 164.8 | 0.563 | 4.130 | 0.392 | 641.0 | 655.7 |
| 45.0 | 1.0000 | 208.4 | 185.4 | 0.629 | 4.132 | 0.424 | 647.9 | 599.8 |
| 50.0 | 1.0022 | 229.1 | 206.0 | 0.693 | 4.133 | 0.455 | 654.2 | 551.4 |
| 60.0 | 1.0071 | 270.4 | 247.2 | 0.819 | 4.135 | 0.512 | 665.3 | 471.9 |
| 70.0 | 1.0125 | 311.8 | 288.5 | 0.942 | 4.140 | 0.566 | 674.4 | 409.9 |
| 80.0 | 1.0185 | 353.2 | 329.8 | 1.061 | 4.147 | 0.617 | 681.6 | 360.6 |
| 90.0 | 1.0251 | 394.7 | 371.1 | 1.176 | 4.156 | 0.666 | 687.3 | 320.7 |
| 100.0 | 1.0322 | 436.3 | 412.6 | 1.290 | 4.167 | 0.714 | 691.6 | 288.0 |
| 125.0 | 1.0523 | 540.9 | 516.7 | 1.561 | 4.202 | 0.832 | 697.3 | 228.1 |
| 150.0 | 1.0761 | 646.5 | 621.8 | 1.818 | 4.245 | 0.953 | 697.1 | 188.2 |
| 175.0 | 1.1038 | 753.3 | 727.9 | 2.063 | 4.302 | 1.085 | 691.9 | 160.1 |
| 200.0 | 1.1362 | 861.8 | 835.7 | 2.299 | 4.380 | 1.238 | 681.6 | 139.6 |
| 225.0 | 1.1746 | 972.6 | 945.6 | 2.527 | 4.490 | 1.425 | 665.9 | 123.8 |
| 250.0 | 1.2207 | 1086 | 1058 | 2.750 | 4.647 | 1.667 | 644.1 | 111.2 |
| 275.0 | 1.2777 | 1205 | 1176 | 2.972 | 4.876 | 2.002 | 615.0 | 100.6 |
| 300.0 | 1.3512 | 1331 | 1300 | 3.197 | 5.231 | 2.515 | 577.1 | 91.09 |
| 325.0 | 1.4536 | 1469 | 1435 | 3.432 | 5.855 | 3.433 | 530.1 | 81.85 |
| 350.0 | 1.6212 | 1630 | 1593 | 3.696 | 7.347 | 5.745 | 474.6 | 71.55 |
| 375.0 | 2.2164 | 1913 | 1862 | 4.140 | 29.44 | 45.00 | 406.1 | 52.24 |
| 400.0 | 7.4787 | 2689 | 2517 | 5.324 | 9.277 | 11.22 | 134.8 | 27.44 |
| 450.0 | 10.408 | 2996 | 2757 | 5.766 | 4.567 | 4.349 | 100.8 | 28.43 |
| 500.0 | 12.394 | 3194 | 2909 | 6.032 | 3.538 | 2.874 | 95.80 | 30.29 |
| 550.0 | 14.050 | 3359 | 3036 | 6.238 | 3.107 | 2.212 | 97.29 | 32.27 |
| 600.0 | 15.532 | 3509 | 3151 | 6.415 | 2.886 | 1.829 | 101.4 | 34.25 |
| 650.0 | 16.906 | 3649 | 3261 | 6.572 | 2.761 | 1.576 | 106.6 | 36.20 |
| 700.0 | 18.205 | 3786 | 3367 | 6.715 | 2.688 | 1.396 | 112.3 | 38.12 |
| 750.0 | 19.452 | 3919 | 3471 | 6.849 | 2.645 | 1.259 | 118.2 | 39.99 |
| 800.0 | 20.658 | 4050 | 3575 | 6.974 | 2.622 | 1.151 | 124.0 | 41.83 |
| 850.0 | 21.833 | 4181 | 3679 | 7.093 | 2.612 | 1.064 | 129.8 | 43.63 |
| 900.0 | 22.983 | 4312 | 3783 | 7.207 | 2.611 | 0.992 | 135.6 | 45.39 |
| 950.0 | 24.113 | 4442 | 3888 | 7.316 | 2.617 | 0.930 | 141.3 | 47.12 |
| 1000.0 | 25.228 | 4573 | 3993 | 7.421 | 2.626 | 0.878 | 146.9 | 48.81 |

Table 1 (cont.)

$$p = 24.0\,\text{MPa} = 240\;\text{bar}$$

| $t$ | $v$ | $h$ | $u$ | $s$ | $c_p$ | $\alpha_p$ | $\lambda$ | $\eta$ |
|---|---|---|---|---|---|---|---|---|
| °C | dm³/kg | kJ/kg | kJ/kg | kJ/kg K | kJ/kg K | $10^{-3}$/K | mW/K m | $\mu$Pa s |
| 0.0 | 0.9885 | 24.0 | 0.3 | 0.0006 | 4.113 | 0.011 | 574.2 | 1739 |
| 5.0 | 0.9887 | 44.6 | 20.8 | 0.075 | 4.102 | 0.082 | 582.8 | 1484 |
| 10.0 | 0.9893 | 65.1 | 41.3 | 0.148 | 4.103 | 0.143 | 591.7 | 1283 |
| 15.0 | 0.9901 | 85.6 | 61.8 | 0.220 | 4.109 | 0.195 | 600.6 | 1123 |
| 20.0 | 0.9912 | 106.2 | 82.4 | 0.291 | 4.115 | 0.242 | 609.4 | 993.7 |
| 25.0 | 0.9925 | 126.8 | 102.9 | 0.360 | 4.120 | 0.284 | 618.0 | 886.3 |
| 30.0 | 0.9940 | 147.4 | 123.5 | 0.429 | 4.124 | 0.323 | 626.3 | 796.5 |
| 35.0 | 0.9957 | 168.0 | 144.1 | 0.496 | 4.127 | 0.359 | 634.1 | 720.6 |
| 40.0 | 0.9976 | 188.6 | 164.7 | 0.563 | 4.128 | 0.392 | 641.5 | 655.8 |
| 45.0 | 0.9996 | 209.3 | 185.3 | 0.628 | 4.130 | 0.424 | 648.3 | 600.0 |
| 50.0 | 1.0018 | 229.9 | 205.9 | 0.693 | 4.131 | 0.455 | 654.7 | 551.6 |
| 60.0 | 1.0067 | 271.2 | 247.1 | 0.819 | 4.133 | 0.512 | 665.8 | 472.2 |
| 70.0 | 1.0121 | 312.6 | 288.3 | 0.941 | 4.138 | 0.565 | 674.8 | 410.2 |
| 80.0 | 1.0181 | 354.0 | 329.6 | 1.060 | 4.145 | 0.616 | 682.1 | 360.9 |
| 90.0 | 1.0246 | 395.5 | 370.9 | 1.176 | 4.154 | 0.665 | 687.8 | 321.0 |
| 100.0 | 1.0317 | 437.1 | 412.3 | 1.289 | 4.165 | 0.713 | 692.1 | 288.3 |
| 125.0 | 1.0518 | 541.6 | 516.4 | 1.560 | 4.200 | 0.830 | 697.8 | 228.3 |
| 150.0 | 1.0755 | 647.2 | 621.3 | 1.817 | 4.243 | 0.950 | 697.8 | 188.4 |
| 175.0 | 1.1031 | 753.9 | 727.4 | 2.062 | 4.299 | 1.081 | 692.6 | 160.4 |
| 200.0 | 1.1354 | 862.3 | 835.0 | 2.297 | 4.376 | 1.232 | 682.4 | 139.8 |
| 225.0 | 1.1735 | 972.9 | 944.8 | 2.525 | 4.484 | 1.417 | 666.9 | 124.1 |
| 250.0 | 1.2193 | 1086 | 1057 | 2.748 | 4.638 | 1.655 | 645.2 | 111.5 |
| 275.0 | 1.2757 | 1205 | 1174 | 2.970 | 4.862 | 1.983 | 616.4 | 100.9 |
| 300.0 | 1.3482 | 1330 | 1298 | 3.194 | 5.206 | 2.480 | 578.9 | 91.41 |
| 325.0 | 1.4485 | 1467 | 1432 | 3.427 | 5.801 | 3.353 | 532.5 | 82.26 |
| 350.0 | 1.6092 | 1626 | 1588 | 3.687 | 7.159 | 5.444 | 478.0 | 72.19 |
| 375.0 | 2.0614 | 1872 | 1823 | 4.073 | 17.40 | 23.09 | 405.5 | 55.94 |
| 400.0 | 6.7315 | 2637 | 2475 | 5.237 | 10.93 | 13.71 | 149.7 | 28.19 |
| 450.0 | 9.7632 | 2973 | 2739 | 5.721 | 4.795 | 4.641 | 104.6 | 28.68 |
| 500.0 | 11.733 | 3179 | 2898 | 5.996 | 3.635 | 2.988 | 97.98 | 30.46 |
| 550.0 | 13.356 | 3348 | 3027 | 6.208 | 3.164 | 2.272 | 98.90 | 32.41 |
| 600.0 | 14.799 | 3499 | 3144 | 6.387 | 2.923 | 1.866 | 102.7 | 34.37 |
| 650.0 | 16.133 | 3642 | 3255 | 6.545 | 2.788 | 1.602 | 107.8 | 36.31 |
| 700.0 | 17.392 | 3779 | 3362 | 6.690 | 2.708 | 1.414 | 113.5 | 38.22 |
| 750.0 | 18.597 | 3913 | 3467 | 6.825 | 2.661 | 1.272 | 119.2 | 40.09 |
| 800.0 | 19.762 | 4045 | 3571 | 6.951 | 2.635 | 1.162 | 125.0 | 41.92 |
| 850.0 | 20.895 | 4177 | 3675 | 7.071 | 2.623 | 1.072 | 130.8 | 43.71 |
| 900.0 | 22.004 | 4308 | 3780 | 7.185 | 2.620 | 0.998 | 136.4 | 45.47 |
| 950.0 | 23.093 | 4439 | 3885 | 7.294 | 2.625 | 0.935 | 142.0 | 47.19 |
| 1000.0 | 24.165 | 4570 | 3991 | 7.399 | 2.633 | 0.882 | 147.6 | 48.88 |

Table 1 (cont.)

$$p = 25.0 \text{ MPa} = 250 \text{ bar}$$

| $t$ | $v$ | $h$ | $u$ | $s$ | $c_p$ | $\alpha_p$ | $\lambda$ | $\eta$ |
|---|---|---|---|---|---|---|---|---|
| °C | dm³/kg | kJ/kg | kJ/kg | kJ/kg K | kJ/kg K | $10^{-3}$/K | mW/K m | $\mu$Pa s |
| 0.0 | 0.9880 | 25.0 | 0.3 | 0.0006 | 4.109 | 0.015 | 574.8 | 1737 |
| 5.0 | 0.9883 | 45.5 | 20.8 | 0.075 | 4.099 | 0.085 | 583.3 | 1482 |
| 10.0 | 0.9888 | 66.0 | 41.3 | 0.148 | 4.100 | 0.145 | 592.1 | 1283 |
| 15.0 | 0.9897 | 86.5 | 61.8 | 0.220 | 4.106 | 0.197 | 601.1 | 1123 |
| 20.0 | 0.9908 | 107.1 | 82.3 | 0.291 | 4.112 | 0.243 | 609.9 | 993.4 |
| 25.0 | 0.9921 | 127.7 | 102.9 | 0.360 | 4.118 | 0.285 | 618.5 | 886.2 |
| 30.0 | 0.9936 | 148.3 | 123.4 | 0.429 | 4.122 | 0.324 | 626.7 | 796.5 |
| 35.0 | 0.9953 | 168.9 | 144.0 | 0.496 | 4.124 | 0.359 | 634.6 | 720.7 |
| 40.0 | 0.9972 | 189.5 | 164.6 | 0.563 | 4.126 | 0.393 | 641.9 | 655.9 |
| 45.0 | 0.9992 | 210.1 | 185.2 | 0.628 | 4.128 | 0.424 | 648.8 | 600.2 |
| 50.0 | 1.0014 | 230.8 | 205.7 | 0.692 | 4.129 | 0.454 | 655.1 | 551.8 |
| 60.0 | 1.0063 | 272.1 | 246.9 | 0.818 | 4.131 | 0.511 | 666.2 | 472.4 |
| 70.0 | 1.0117 | 313.4 | 288.1 | 0.940 | 4.136 | 0.565 | 675.3 | 410.4 |
| 80.0 | 1.0177 | 354.8 | 329.4 | 1.059 | 4.143 | 0.615 | 682.6 | 361.1 |
| 90.0 | 1.0242 | 396.3 | 370.7 | 1.175 | 4.152 | 0.664 | 688.3 | 321.2 |
| 100.0 | 1.0313 | 437.9 | 412.1 | 1.288 | 4.163 | 0.711 | 692.7 | 288.5 |
| 125.0 | 1.0513 | 542.3 | 516.1 | 1.559 | 4.197 | 0.828 | 698.4 | 228.6 |
| 150.0 | 1.0749 | 647.8 | 620.9 | 1.816 | 4.240 | 0.947 | 698.4 | 188.6 |
| 175.0 | 1.1024 | 754.5 | 726.9 | 2.061 | 4.296 | 1.077 | 693.3 | 160.6 |
| 200.0 | 1.1345 | 862.7 | 834.4 | 2.296 | 4.371 | 1.227 | 683.2 | 140.0 |
| 225.0 | 1.1724 | 973.3 | 944.0 | 2.524 | 4.478 | 1.409 | 667.8 | 124.3 |
| 250.0 | 1.2179 | 1087 | 1056 | 2.746 | 4.629 | 1.643 | 646.3 | 111.7 |
| 275.0 | 1.2737 | 1205 | 1173 | 2.967 | 4.848 | 1.965 | 617.7 | 101.1 |
| 300.0 | 1.3453 | 1330 | 1296 | 3.190 | 5.182 | 2.446 | 580.7 | 91.72 |
| 325.0 | 1.4435 | 1466 | 1430 | 3.422 | 5.751 | 3.278 | 534.8 | 82.65 |
| 350.0 | 1.5981 | 1623 | 1583 | 3.679 | 6.994 | 5.184 | 481.3 | 72.81 |
| 375.0 | 1.9794 | 1849 | 1799 | 4.034 | 13.72 | 16.54 | 411.2 | 58.22 |
| 400.0 | 6.0014 | 2578 | 2428 | 5.139 | 13.27 | 17.30 | 169.0 | 29.18 |
| 450.0 | 9.1666 | 2950 | 2721 | 5.676 | 5.043 | 4.960 | 108.8 | 28.96 |
| 500.0 | 11.123 | 3164 | 2886 | 5.962 | 3.737 | 3.107 | 100.3 | 30.64 |
| 550.0 | 12.716 | 3336 | 3018 | 6.178 | 3.221 | 2.335 | 100.6 | 32.55 |
| 600.0 | 14.126 | 3490 | 3137 | 6.359 | 2.961 | 1.905 | 104.1 | 34.49 |
| 650.0 | 15.423 | 3634 | 3248 | 6.520 | 2.815 | 1.627 | 109.1 | 36.42 |
| 700.0 | 16.644 | 3772 | 3356 | 6.666 | 2.729 | 1.432 | 114.6 | 38.32 |
| 750.0 | 17.812 | 3908 | 3462 | 6.801 | 2.678 | 1.286 | 120.3 | 40.18 |
| 800.0 | 18.938 | 4041 | 3567 | 6.928 | 2.649 | 1.172 | 126.0 | 42.00 |
| 850.0 | 20.033 | 4173 | 3672 | 7.048 | 2.634 | 1.080 | 131.7 | 43.79 |
| 900.0 | 21.104 | 4304 | 3777 | 7.163 | 2.630 | 1.004 | 137.3 | 45.54 |
| 950.0 | 22.154 | 4436 | 3882 | 7.273 | 2.632 | 0.940 | 142.8 | 47.26 |
| 1000.0 | 23.188 | 4568 | 3988 | 7.379 | 2.640 | 0.886 | 148.3 | 48.95 |

Table 1 (cont.) $p = 27.5\,\text{MPa} = 275\ \text{bar}$

| $t$ | $v$ | $h$ | $u$ | $s$ | $c_p$ | $\alpha_p$ | $\lambda$ | $\eta$ |
|------|--------|-------|-------|--------|--------|-----------|--------|--------|
| °C | dm³/kg | kJ/kg | kJ/kg | kJ/kg K | kJ/kg K | 10⁻³/K | mW/K m | µPa s |
| 0.0 | 0.9868 | 27.5 | 0.3 | 0.0006 | 4.099 | 0.024 | 576.1 | 1732 |
| 5.0 | 0.9871 | 47.9 | 20.8 | 0.075 | 4.090 | 0.092 | 584.6 | 1479 |
| 10.0 | 0.9877 | 68.4 | 41.2 | 0.148 | 4.092 | 0.150 | 593.4 | 1281 |
| 15.0 | 0.9886 | 88.9 | 61.7 | 0.219 | 4.099 | 0.201 | 602.2 | 1122 |
| 20.0 | 0.9897 | 109.4 | 82.2 | 0.290 | 4.106 | 0.247 | 611.0 | 992.8 |
| 25.0 | 0.9910 | 129.9 | 102.7 | 0.359 | 4.112 | 0.288 | 619.6 | 885.9 |
| 30.0 | 0.9926 | 150.5 | 123.2 | 0.428 | 4.116 | 0.325 | 627.9 | 796.5 |
| 35.0 | 0.9943 | 171.1 | 143.7 | 0.495 | 4.119 | 0.360 | 635.7 | 720.9 |
| 40.0 | 0.9961 | 191.7 | 164.3 | 0.562 | 4.121 | 0.393 | 643.1 | 656.3 |
| 45.0 | 0.9982 | 212.3 | 184.8 | 0.627 | 4.123 | 0.425 | 649.9 | 600.6 |
| 50.0 | 1.0004 | 232.9 | 205.4 | 0.691 | 4.124 | 0.454 | 656.3 | 552.3 |
| 60.0 | 1.0052 | 274.2 | 246.5 | 0.817 | 4.126 | 0.510 | 667.4 | 473.0 |
| 70.0 | 1.0106 | 315.4 | 287.7 | 0.939 | 4.131 | 0.563 | 676.6 | 411.1 |
| 80.0 | 1.0166 | 356.8 | 328.8 | 1.058 | 4.138 | 0.613 | 683.9 | 361.8 |
| 90.0 | 1.0231 | 398.2 | 370.1 | 1.173 | 4.147 | 0.661 | 689.6 | 321.9 |
| 100.0 | 1.0301 | 439.7 | 411.4 | 1.286 | 4.158 | 0.708 | 694.0 | 289.2 |
| 125.0 | 1.0500 | 544.1 | 515.2 | 1.557 | 4.192 | 0.823 | 699.9 | 229.2 |
| 150.0 | 1.0734 | 649.4 | 619.9 | 1.813 | 4.234 | 0.940 | 700.1 | 189.2 |
| 175.0 | 1.1006 | 755.9 | 725.6 | 2.058 | 4.287 | 1.067 | 695.2 | 161.2 |
| 200.0 | 1.1324 | 863.9 | 832.8 | 2.292 | 4.361 | 1.214 | 685.3 | 140.6 |
| 225.0 | 1.1698 | 974.2 | 942.0 | 2.519 | 4.464 | 1.390 | 670.1 | 124.9 |
| 250.0 | 1.2144 | 1087 | 1054 | 2.741 | 4.608 | 1.615 | 649.1 | 112.3 |
| 275.0 | 1.2690 | 1205 | 1170 | 2.961 | 4.816 | 1.920 | 621.1 | 101.8 |
| 300.0 | 1.3383 | 1329 | 1292 | 3.182 | 5.127 | 2.367 | 585.0 | 92.48 |
| 325.0 | 1.4319 | 1463 | 1423 | 3.411 | 5.637 | 3.110 | 540.5 | 83.61 |
| 350.0 | 1.5734 | 1614 | 1571 | 3.659 | 6.660 | 4.660 | 489.1 | 74.23 |
| 375.0 | 1.8621 | 1813 | 1762 | 3.972 | 10.22 | 10.48 | 425.7 | 61.97 |
| 400.0 | 4.1803 | 2378 | 2263 | 4.824 | 24.84 | 36.16 | 248.4 | 33.97 |
| 450.0 | 7.8506 | 2888 | 2673 | 5.561 | 5.764 | 5.900 | 121.0 | 29.78 |
| 500.0 | 9.7900 | 3124 | 2855 | 5.877 | 4.010 | 3.428 | 106.6 | 31.15 |
| 550.0 | 11.321 | 3306 | 2995 | 6.105 | 3.371 | 2.498 | 105.0 | 32.94 |
| 600.0 | 12.655 | 3466 | 3118 | 6.294 | 3.059 | 2.003 | 107.8 | 34.82 |
| 650.0 | 13.873 | 3615 | 3233 | 6.459 | 2.885 | 1.693 | 112.4 | 36.71 |
| 700.0 | 15.013 | 3756 | 3343 | 6.608 | 2.782 | 1.478 | 117.6 | 38.58 |
| 750.0 | 16.098 | 3893 | 3451 | 6.746 | 2.719 | 1.320 | 123.1 | 40.42 |
| 800.0 | 17.141 | 4028 | 3557 | 6.875 | 2.682 | 1.197 | 128.6 | 42.23 |
| 850.0 | 18.153 | 4162 | 3663 | 6.997 | 2.662 | 1.100 | 134.1 | 44.00 |
| 900.0 | 19.140 | 4295 | 3768 | 7.112 | 2.653 | 1.020 | 139.5 | 45.74 |
| 950.0 | 20.106 | 4427 | 3874 | 7.223 | 2.652 | 0.953 | 144.8 | 47.45 |
| 1000.0 | 21.056 | 4560 | 3981 | 7.329 | 2.657 | 0.896 | 150.0 | 49.12 |

Table 1 (cont.) $p = 30.0\,\text{MPa} = 300\ \text{bar}$

| $t$ | $v$ | $h$ | $u$ | $s$ | $c_p$ | $\alpha_p$ | $\lambda$ | $\eta$ |
|---|---|---|---|---|---|---|---|---|
| °C | dm³/kg | kJ/kg | kJ/kg | kJ/kg K | kJ/kg K | $10^{-3}$/K | mW/K m | μPa s |
| 0.0 | 0.9857 | 29.9 | 0.4 | 0.0005 | 4.089 | 0.032 | 577.5 | 1728 |
| 5.0 | 0.9860 | 50.3 | 20.8 | 0.075 | 4.081 | 0.098 | 585.8 | 1476 |
| 10.0 | 0.9866 | 70.8 | 41.2 | 0.147 | 4.085 | 0.155 | 594.6 | 1279 |
| 15.0 | 0.9875 | 91.2 | 61.6 | 0.219 | 4.092 | 0.205 | 603.4 | 1121 |
| 20.0 | 0.9887 | 111.7 | 82.0 | 0.289 | 4.100 | 0.250 | 612.2 | 992.2 |
| 25.0 | 0.9900 | 132.2 | 102.5 | 0.359 | 4.106 | 0.290 | 620.8 | 885.7 |
| 30.0 | 0.9915 | 152.7 | 123.0 | 0.427 | 4.111 | 0.327 | 629.0 | 796.5 |
| 35.0 | 0.9932 | 173.3 | 143.5 | 0.494 | 4.114 | 0.362 | 636.8 | 721.1 |
| 40.0 | 0.9951 | 193.9 | 164.0 | 0.561 | 4.116 | 0.394 | 644.2 | 656.6 |
| 45.0 | 0.9971 | 214.5 | 184.5 | 0.626 | 4.118 | 0.425 | 651.1 | 601.1 |
| 50.0 | 0.9993 | 235.0 | 205.1 | 0.690 | 4.119 | 0.454 | 657.4 | 552.9 |
| 60.0 | 1.0042 | 276.2 | 246.1 | 0.816 | 4.122 | 0.509 | 668.6 | 473.7 |
| 70.0 | 1.0096 | 317.5 | 287.2 | 0.938 | 4.126 | 0.561 | 677.8 | 411.8 |
| 80.0 | 1.0155 | 358.8 | 328.3 | 1.056 | 4.133 | 0.611 | 685.1 | 362.5 |
| 90.0 | 1.0220 | 400.2 | 369.5 | 1.172 | 4.143 | 0.658 | 690.9 | 322.6 |
| 100.0 | 1.0290 | 441.6 | 410.8 | 1.284 | 4.154 | 0.705 | 695.3 | 289.9 |
| 125.0 | 1.0487 | 545.9 | 514.4 | 1.555 | 4.186 | 0.818 | 701.4 | 229.9 |
| 150.0 | 1.0719 | 651.0 | 618.9 | 1.811 | 4.227 | 0.933 | 701.7 | 189.9 |
| 175.0 | 1.0989 | 757.3 | 724.4 | 2.055 | 4.279 | 1.058 | 697.0 | 161.8 |
| 200.0 | 1.1303 | 865.2 | 831.2 | 2.289 | 4.350 | 1.201 | 687.3 | 141.2 |
| 225.0 | 1.1672 | 975.1 | 940.1 | 2.515 | 4.449 | 1.372 | 672.4 | 125.5 |
| 250.0 | 1.2110 | 1087 | 1051 | 2.736 | 4.588 | 1.589 | 651.8 | 112.9 |
| 275.0 | 1.2644 | 1204 | 1167 | 2.955 | 4.785 | 1.878 | 624.3 | 102.4 |
| 300.0 | 1.3317 | 1327 | 1288 | 3.174 | 5.075 | 2.295 | 589.1 | 93.22 |
| 325.0 | 1.4211 | 1460 | 1417 | 3.400 | 5.537 | 2.964 | 545.9 | 84.52 |
| 350.0 | 1.5522 | 1608 | 1561 | 3.642 | 6.401 | 4.261 | 496.3 | 75.51 |
| 375.0 | 1.7913 | 1791 | 1737 | 3.930 | 8.768 | 8.044 | 437.9 | 64.58 |
| 400.0 | 2.7929 | 2150 | 2066 | 4.472 | 25.08 | 36.75 | 329.7 | 44.00 |
| 450.0 | 6.7363 | 2822 | 2620 | 5.444 | 6.655 | 7.082 | 136.0 | 30.85 |
| 500.0 | 8.6761 | 3083 | 2823 | 5.794 | 4.312 | 3.784 | 113.7 | 31.73 |
| 550.0 | 10.158 | 3276 | 2972 | 6.036 | 3.530 | 2.670 | 109.8 | 33.37 |
| 600.0 | 11.431 | 3443 | 3100 | 6.232 | 3.160 | 2.105 | 111.7 | 35.17 |
| 650.0 | 12.583 | 3595 | 3218 | 6.402 | 2.956 | 1.760 | 115.8 | 37.02 |
| 700.0 | 13.655 | 3740 | 3330 | 6.555 | 2.836 | 1.525 | 120.7 | 38.86 |
| 750.0 | 14.671 | 3879 | 3439 | 6.695 | 2.761 | 1.354 | 126.0 | 40.68 |
| 800.0 | 15.645 | 4016 | 3547 | 6.825 | 2.716 | 1.223 | 131.3 | 42.47 |
| 850.0 | 16.587 | 4151 | 3654 | 6.948 | 2.690 | 1.120 | 136.5 | 44.22 |
| 900.0 | 17.504 | 4285 | 3760 | 7.065 | 2.677 | 1.035 | 141.7 | 45.95 |
| 950.0 | 18.400 | 4419 | 3867 | 7.177 | 2.672 | 0.965 | 146.8 | 47.64 |
| 1000.0 | 19.281 | 4553 | 3974 | 7.284 | 2.674 | 0.906 | 151.9 | 49.30 |

Table 1 (cont.)

$$p = 32.5\,\text{MPa} = 325\ \text{bar}$$

| $t$ | $v$ | $h$ | $u$ | $s$ | $c_p$ | $\alpha_p$ | $\lambda$ | $\eta$ |
|---|---|---|---|---|---|---|---|---|
| °C | dm³/kg | kJ/kg | kJ/kg | kJ/kg K | kJ/kg K | $10^{-3}$/K | mW/K m | $\mu$Pa s |
| 0.0 | 0.9845 | 32.4 | 0.4 | 0.0004 | 4.079 | 0.041 | 578.8 | 1724 |
| 5.0 | 0.9849 | 52.7 | 20.7 | 0.074 | 4.073 | 0.105 | 587.1 | 1474 |
| 10.0 | 0.9856 | 73.1 | 41.1 | 0.147 | 4.077 | 0.161 | 595.8 | 1277 |
| 15.0 | 0.9865 | 93.5 | 61.5 | 0.218 | 4.085 | 0.209 | 604.6 | 1120 |
| 20.0 | 0.9876 | 114.0 | 81.9 | 0.289 | 4.093 | 0.253 | 613.3 | 991.7 |
| 25.0 | 0.9890 | 134.5 | 102.3 | 0.358 | 4.100 | 0.293 | 621.9 | 885.5 |
| 30.0 | 0.9905 | 155.0 | 122.8 | 0.426 | 4.105 | 0.329 | 630.1 | 796.6 |
| 35.0 | 0.9922 | 175.5 | 143.3 | 0.493 | 4.109 | 0.363 | 638.0 | 721.3 |
| 40.0 | 0.9941 | 196.0 | 163.7 | 0.560 | 4.111 | 0.395 | 645.4 | 657.0 |
| 45.0 | 0.9961 | 216.6 | 184.2 | 0.625 | 4.113 | 0.425 | 652.2 | 601.6 |
| 50.0 | 0.9983 | 237.2 | 204.7 | 0.689 | 4.114 | 0.454 | 658.6 | 553.4 |
| 60.0 | 1.0031 | 278.3 | 245.7 | 0.814 | 4.117 | 0.509 | 669.8 | 474.3 |
| 70.0 | 1.0085 | 319.5 | 286.7 | 0.936 | 4.122 | 0.560 | 679.0 | 412.4 |
| 80.0 | 1.0144 | 360.8 | 327.8 | 1.055 | 4.129 | 0.609 | 686.4 | 363.1 |
| 90.0 | 1.0209 | 402.1 | 368.9 | 1.170 | 4.138 | 0.656 | 692.2 | 323.3 |
| 100.0 | 1.0278 | 443.5 | 410.1 | 1.283 | 4.149 | 0.701 | 696.7 | 290.5 |
| 125.0 | 1.0475 | 547.6 | 513.6 | 1.553 | 4.181 | 0.813 | 702.9 | 230.5 |
| 150.0 | 1.0705 | 652.7 | 617.9 | 1.808 | 4.221 | 0.927 | 703.3 | 190.5 |
| 175.0 | 1.0972 | 758.8 | 723.1 | 2.052 | 4.272 | 1.049 | 698.8 | 162.4 |
| 200.0 | 1.1283 | 866.4 | 829.7 | 2.286 | 4.340 | 1.188 | 689.3 | 141.8 |
| 225.0 | 1.1646 | 976.0 | 938.2 | 2.511 | 4.436 | 1.355 | 674.7 | 126.0 |
| 250.0 | 1.2078 | 1088 | 1049 | 2.732 | 4.569 | 1.563 | 654.4 | 113.5 |
| 275.0 | 1.2600 | 1204 | 1163 | 2.949 | 4.756 | 1.839 | 627.5 | 103.1 |
| 300.0 | 1.3253 | 1327 | 1283 | 3.167 | 5.028 | 2.229 | 593.2 | 93.94 |
| 325.0 | 1.4111 | 1457 | 1411 | 3.390 | 5.449 | 2.837 | 551.0 | 85.39 |
| 350.0 | 1.5335 | 1602 | 1552 | 3.626 | 6.194 | 3.945 | 503.0 | 76.69 |
| 375.0 | 1.7404 | 1774 | 1718 | 3.898 | 7.936 | 6.684 | 448.3 | 66.65 |
| 400.0 | 2.3073 | 2041 | 1966 | 4.301 | 15.54 | 19.73 | 363.0 | 51.39 |
| 450.0 | 5.7810 | 2750 | 2562 | 5.322 | 7.739 | 8.540 | 154.4 | 32.24 |
| 500.0 | 7.7322 | 3041 | 2789 | 5.712 | 4.644 | 4.177 | 121.7 | 32.41 |
| 550.0 | 9.1747 | 3246 | 2947 | 5.970 | 3.699 | 2.853 | 115.1 | 33.84 |
| 600.0 | 10.396 | 3418 | 3081 | 6.174 | 3.265 | 2.210 | 115.9 | 35.56 |
| 650.0 | 11.492 | 3575 | 3202 | 6.348 | 3.029 | 1.827 | 119.4 | 37.35 |
| 700.0 | 12.507 | 3723 | 3317 | 6.504 | 2.890 | 1.572 | 124.0 | 39.15 |
| 750.0 | 13.464 | 3865 | 3428 | 6.647 | 2.804 | 1.388 | 129.0 | 40.95 |
| 800.0 | 14.379 | 4004 | 3537 | 6.779 | 2.751 | 1.249 | 134.1 | 42.72 |
| 850.0 | 15.262 | 4141 | 3645 | 6.904 | 2.718 | 1.139 | 139.1 | 44.45 |
| 900.0 | 16.120 | 4276 | 3752 | 7.021 | 2.700 | 1.051 | 144.1 | 46.16 |
| 950.0 | 16.958 | 4411 | 3860 | 7.134 | 2.692 | 0.977 | 149.0 | 47.84 |
| 1000.0 | 17.778 | 4545 | 3968 | 7.242 | 2.692 | 0.915 | 153.8 | 49.49 |

Table 1 (cont.)

$$p = 35.0 \, \text{MPa} = 350 \ \text{bar}$$

| $t$ | $v$ | $h$ | $u$ | $s$ | $c_p$ | $\alpha_p$ | $\lambda$ | $\eta$ |
|---|---|---|---|---|---|---|---|---|
| °C | dm³/kg | kJ/kg | kJ/kg | kJ/kg K | kJ/kg K | $10^{-3}$/K | mW/K m | $\mu$Pa s |
| 0.0 | 0.9834 | 34.8 | 0.4 | 0.0003 | 4.070 | 0.049 | 580.2 | 1719 |
| 5.0 | 0.9838 | 55.1 | 20.7 | 0.074 | 4.065 | 0.112 | 588.4 | 1471 |
| 10.0 | 0.9845 | 75.5 | 41.0 | 0.147 | 4.070 | 0.166 | 597.0 | 1275 |
| 15.0 | 0.9854 | 95.8 | 61.3 | 0.218 | 4.079 | 0.213 | 605.7 | 1119 |
| 20.0 | 0.9866 | 116.2 | 81.7 | 0.288 | 4.087 | 0.256 | 614.5 | 991.1 |
| 25.0 | 0.9879 | 136.7 | 102.1 | 0.357 | 4.095 | 0.295 | 623.0 | 885.3 |
| 30.0 | 0.9895 | 157.2 | 122.6 | 0.425 | 4.100 | 0.331 | 631.3 | 796.7 |
| 35.0 | 0.9912 | 177.7 | 143.0 | 0.493 | 4.104 | 0.364 | 639.1 | 721.6 |
| 40.0 | 0.9931 | 198.2 | 163.5 | 0.559 | 4.106 | 0.396 | 646.5 | 657.4 |
| 45.0 | 0.9951 | 218.8 | 183.9 | 0.624 | 4.108 | 0.425 | 653.4 | 602.0 |
| 50.0 | 0.9973 | 239.3 | 204.4 | 0.688 | 4.109 | 0.454 | 659.8 | 554.0 |
| 60.0 | 1.0021 | 280.4 | 245.3 | 0.813 | 4.113 | 0.508 | 671.0 | 474.9 |
| 70.0 | 1.0075 | 321.6 | 286.3 | 0.935 | 4.117 | 0.558 | 680.2 | 413.1 |
| 80.0 | 1.0134 | 362.8 | 327.3 | 1.053 | 4.124 | 0.607 | 687.7 | 363.8 |
| 90.0 | 1.0198 | 404.1 | 368.4 | 1.168 | 4.133 | 0.653 | 693.5 | 323.9 |
| 100.0 | 1.0267 | 445.4 | 409.5 | 1.281 | 4.144 | 0.698 | 698.0 | 291.2 |
| 125.0 | 1.0462 | 549.4 | 512.8 | 1.550 | 4.176 | 0.808 | 704.3 | 231.1 |
| 150.0 | 1.0690 | 654.3 | 616.9 | 1.806 | 4.215 | 0.920 | 704.9 | 191.1 |
| 175.0 | 1.0955 | 760.2 | 721.9 | 2.049 | 4.264 | 1.040 | 700.6 | 162.9 |
| 200.0 | 1.1263 | 867.6 | 828.2 | 2.282 | 4.331 | 1.176 | 691.3 | 142.3 |
| 225.0 | 1.1621 | 977.0 | 936.3 | 2.508 | 4.423 | 1.338 | 677.0 | 126.6 |
| 250.0 | 1.2046 | 1089 | 1046 | 2.727 | 4.551 | 1.539 | 657.0 | 114.1 |
| 275.0 | 1.2557 | 1204 | 1160 | 2.943 | 4.729 | 1.802 | 630.7 | 103.7 |
| 300.0 | 1.3193 | 1326 | 1279 | 3.160 | 4.984 | 2.169 | 597.1 | 94.64 |
| 325.0 | 1.4017 | 1455 | 1406 | 3.380 | 5.371 | 2.724 | 556.0 | 86.22 |
| 350.0 | 1.5168 | 1596 | 1543 | 3.612 | 6.022 | 3.687 | 509.3 | 77.79 |
| 375.0 | 1.7007 | 1761 | 1702 | 3.871 | 7.384 | 5.799 | 457.4 | 68.40 |
| 400.0 | 2.1057 | 1988 | 1914 | 4.214 | 11.67 | 12.96 | 384.4 | 55.79 |
| 450.0 | 4.9596 | 2672 | 2499 | 5.197 | 8.982 | 10.22 | 176.4 | 34.03 |
| 500.0 | 6.9235 | 2997 | 2754 | 5.632 | 5.006 | 4.605 | 130.7 | 33.19 |
| 550.0 | 8.3332 | 3214 | 2923 | 5.905 | 3.875 | 3.043 | 120.8 | 34.37 |
| 600.0 | 9.5106 | 3394 | 3061 | 6.117 | 3.373 | 2.318 | 120.3 | 35.97 |
| 650.0 | 10.558 | 3556 | 3186 | 6.297 | 3.104 | 1.896 | 123.3 | 37.70 |
| 700.0 | 11.523 | 3706 | 3303 | 6.456 | 2.945 | 1.619 | 127.5 | 39.47 |
| 750.0 | 12.431 | 3851 | 3416 | 6.601 | 2.847 | 1.422 | 132.2 | 41.23 |
| 800.0 | 13.295 | 3992 | 3526 | 6.735 | 2.785 | 1.274 | 137.0 | 42.97 |
| 850.0 | 14.128 | 4130 | 3635 | 6.861 | 2.747 | 1.159 | 141.8 | 44.69 |
| 900.0 | 14.935 | 4267 | 3744 | 6.980 | 2.724 | 1.066 | 146.5 | 46.38 |
| 950.0 | 15.721 | 4403 | 3852 | 7.094 | 2.713 | 0.990 | 151.2 | 48.05 |
| 1000.0 | 16.491 | 4538 | 3961 | 7.202 | 2.709 | 0.925 | 155.8 | 49.68 |

Table 1 (cont.)

$$p = 37.5\,\text{MPa} = 375\ \text{bar}$$

| $t$ | $v$ | $h$ | $u$ | $s$ | $c_p$ | $\alpha_p$ | $\lambda$ | $\eta$ |
|---|---|---|---|---|---|---|---|---|
| °C | dm³/kg | kJ/kg | kJ/kg | kJ/kg K | kJ/kg K | $10^{-3}$/K | mW/K m | $\mu$Pa s |
| 0.0 | 0.9823 | 37.2 | 0.4 | 0.0002 | 4.061 | 0.057 | 581.5 | 1715 |
| 5.0 | 0.9827 | 57.5 | 20.7 | 0.074 | 4.057 | 0.118 | 589.6 | 1468 |
| 10.0 | 0.9834 | 77.8 | 40.9 | 0.146 | 4.063 | 0.171 | 598.2 | 1274 |
| 15.0 | 0.9844 | 98.1 | 61.2 | 0.217 | 4.072 | 0.217 | 606.9 | 1118 |
| 20.0 | 0.9855 | 118.5 | 81.6 | 0.287 | 4.082 | 0.259 | 615.6 | 990.7 |
| 25.0 | 0.9869 | 139.0 | 101.9 | 0.357 | 4.089 | 0.297 | 624.1 | 885.2 |
| 30.0 | 0.9885 | 159.4 | 122.3 | 0.425 | 4.095 | 0.333 | 632.4 | 796.8 |
| 35.0 | 0.9902 | 179.9 | 142.8 | 0.492 | 4.099 | 0.365 | 640.2 | 721.9 |
| 40.0 | 0.9921 | 200.4 | 163.2 | 0.558 | 4.101 | 0.396 | 647.6 | 657.8 |
| 45.0 | 0.9941 | 220.9 | 183.6 | 0.623 | 4.103 | 0.426 | 654.5 | 602.5 |
| 50.0 | 0.9963 | 241.4 | 204.1 | 0.687 | 4.105 | 0.454 | 660.9 | 554.5 |
| 60.0 | 1.0011 | 282.5 | 245.0 | 0.812 | 4.108 | 0.507 | 672.1 | 475.5 |
| 70.0 | 1.0065 | 323.6 | 285.9 | 0.933 | 4.113 | 0.557 | 681.4 | 413.8 |
| 80.0 | 1.0123 | 364.8 | 326.8 | 1.052 | 4.120 | 0.605 | 688.9 | 364.5 |
| 90.0 | 1.0187 | 406.0 | 367.8 | 1.167 | 4.129 | 0.650 | 694.8 | 324.6 |
| 100.0 | 1.0256 | 447.3 | 408.9 | 1.279 | 4.139 | 0.695 | 699.3 | 291.9 |
| 125.0 | 1.0450 | 551.2 | 512.0 | 1.548 | 4.171 | 0.804 | 705.8 | 231.8 |
| 150.0 | 1.0676 | 655.9 | 615.9 | 1.803 | 4.208 | 0.914 | 706.6 | 191.7 |
| 175.0 | 1.0939 | 761.7 | 720.7 | 2.046 | 4.256 | 1.031 | 702.4 | 163.5 |
| 200.0 | 1.1243 | 868.9 | 826.7 | 2.279 | 4.321 | 1.164 | 693.3 | 142.9 |
| 225.0 | 1.1597 | 978.0 | 934.5 | 2.504 | 4.410 | 1.322 | 679.2 | 127.2 |
| 250.0 | 1.2015 | 1089 | 1044 | 2.722 | 4.533 | 1.516 | 659.6 | 114.7 |
| 275.0 | 1.2516 | 1204 | 1158 | 2.938 | 4.703 | 1.767 | 633.8 | 104.3 |
| 300.0 | 1.3135 | 1325 | 1276 | 3.153 | 4.944 | 2.112 | 600.9 | 95.32 |
| 325.0 | 1.3929 | 1453 | 1400 | 3.371 | 5.300 | 2.623 | 560.8 | 87.02 |
| 350.0 | 1.5017 | 1592 | 1535 | 3.598 | 5.877 | 3.471 | 515.3 | 78.82 |
| 375.0 | 1.6682 | 1751 | 1688 | 3.848 | 6.984 | 5.169 | 465.6 | 69.94 |
| 400.0 | 1.9892 | 1954 | 1880 | 4.156 | 9.806 | 9.783 | 400.8 | 58.89 |
| 450.0 | 4.2626 | 2592 | 2432 | 5.070 | 10.21 | 11.87 | 201.4 | 36.29 |
| 500.0 | 6.2251 | 2952 | 2718 | 5.553 | 5.393 | 5.062 | 140.6 | 34.09 |
| 550.0 | 7.6061 | 3183 | 2898 | 5.843 | 4.060 | 3.241 | 126.9 | 34.95 |
| 600.0 | 8.7449 | 3370 | 3042 | 6.063 | 3.484 | 2.427 | 125.0 | 36.42 |
| 650.0 | 9.7509 | 3536 | 3170 | 6.248 | 3.180 | 1.966 | 127.3 | 38.08 |
| 700.0 | 10.673 | 3690 | 3290 | 6.411 | 3.001 | 1.667 | 131.1 | 39.79 |
| 750.0 | 11.536 | 3837 | 3404 | 6.558 | 2.890 | 1.456 | 135.4 | 41.52 |
| 800.0 | 12.357 | 3980 | 3516 | 6.694 | 2.820 | 1.300 | 140.0 | 43.24 |
| 850.0 | 13.145 | 4119 | 3626 | 6.822 | 2.775 | 1.178 | 144.6 | 44.94 |
| 900.0 | 13.908 | 4257 | 3736 | 6.942 | 2.748 | 1.081 | 149.1 | 46.61 |
| 950.0 | 14.650 | 4394 | 3845 | 7.056 | 2.733 | 1.002 | 153.5 | 48.26 |
| 1000.0 | 15.376 | 4531 | 3954 | 7.165 | 2.726 | 0.935 | 157.8 | 49.88 |

Table 1 (cont.) $p = 40.0\,\mathrm{MPa} = 400\ \mathrm{bar}$

| $t$ | $v$ | $h$ | $u$ | $s$ | $c_p$ | $\alpha_p$ | $\lambda$ | $\eta$ |
|---|---|---|---|---|---|---|---|---|
| °C | dm³/kg | kJ/kg | kJ/kg | kJ/kg K | kJ/kg K | $10^{-3}$/K | mW/K m | µPa s |
| 0.0 | 0.9811 | 39.6 | 0.4 | 0.00003 | 4.053 | 0.065 | 582.9 | 1711 |
| 5.0 | 0.9816 | 59.9 | 20.6 | 0.073 | 4.050 | 0.124 | 590.9 | 1466 |
| 10.0 | 0.9823 | 80.1 | 40.8 | 0.146 | 4.056 | 0.176 | 599.4 | 1272 |
| 15.0 | 0.9833 | 100.4 | 61.1 | 0.217 | 4.066 | 0.221 | 608.1 | 1117 |
| 20.0 | 0.9845 | 120.8 | 81.4 | 0.287 | 4.076 | 0.262 | 616.7 | 990.2 |
| 25.0 | 0.9859 | 141.2 | 101.8 | 0.356 | 4.084 | 0.300 | 625.3 | 885.1 |
| 30.0 | 0.9875 | 161.6 | 122.1 | 0.424 | 4.090 | 0.334 | 633.5 | 796.9 |
| 35.0 | 0.9892 | 182.1 | 142.5 | 0.491 | 4.094 | 0.367 | 641.4 | 722.2 |
| 40.0 | 0.9911 | 202.6 | 162.9 | 0.557 | 4.097 | 0.397 | 648.8 | 658.2 |
| 45.0 | 0.9931 | 223.1 | 183.3 | 0.622 | 4.099 | 0.426 | 655.7 | 603.0 |
| 50.0 | 0.9953 | 243.6 | 203.7 | 0.685 | 4.100 | 0.454 | 662.1 | 555.1 |
| 60.0 | 1.0001 | 284.6 | 244.6 | 0.810 | 4.104 | 0.506 | 673.3 | 476.2 |
| 70.0 | 1.0054 | 325.6 | 285.4 | 0.932 | 4.108 | 0.556 | 682.6 | 414.4 |
| 80.0 | 1.0113 | 366.7 | 326.3 | 1.050 | 4.115 | 0.603 | 690.1 | 365.2 |
| 90.0 | 1.0176 | 407.9 | 367.2 | 1.165 | 4.124 | 0.648 | 696.1 | 325.3 |
| 100.0 | 1.0245 | 449.2 | 408.3 | 1.277 | 4.135 | 0.692 | 700.7 | 292.5 |
| 125.0 | 1.0437 | 553.0 | 511.2 | 1.546 | 4.165 | 0.799 | 707.3 | 232.4 |
| 150.0 | 1.0662 | 657.6 | 614.9 | 1.801 | 4.202 | 0.907 | 708.2 | 192.3 |
| 175.0 | 1.0923 | 763.2 | 719.5 | 2.044 | 4.249 | 1.023 | 704.1 | 164.1 |
| 200.0 | 1.1223 | 870.1 | 825.3 | 2.276 | 4.312 | 1.153 | 695.3 | 143.5 |
| 225.0 | 1.1573 | 979.0 | 932.7 | 2.500 | 4.398 | 1.306 | 681.5 | 127.7 |
| 250.0 | 1.1984 | 1090 | 1042 | 2.718 | 4.516 | 1.494 | 662.2 | 115.3 |
| 275.0 | 1.2476 | 1205 | 1155 | 2.932 | 4.679 | 1.735 | 636.8 | 104.9 |
| 300.0 | 1.3079 | 1324 | 1272 | 3.146 | 4.906 | 2.060 | 604.6 | 95.99 |
| 325.0 | 1.3846 | 1451 | 1395 | 3.362 | 5.235 | 2.532 | 565.4 | 87.80 |
| 350.0 | 1.4879 | 1588 | 1528 | 3.586 | 5.752 | 3.287 | 521.0 | 79.80 |
| 375.0 | 1.6405 | 1742 | 1676 | 3.828 | 6.678 | 4.693 | 473.1 | 71.32 |
| 400.0 | 1.9096 | 1930 | 1854 | 4.113 | 8.717 | 7.972 | 414.0 | 61.31 |
| 450.0 | 3.6912 | 2513 | 2365 | 4.946 | 11.08 | 12.93 | 227.7 | 39.02 |
| 500.0 | 5.6188 | 2906 | 2681 | 5.474 | 5.799 | 5.536 | 151.6 | 35.12 |
| 550.0 | 6.9727 | 3151 | 2872 | 5.782 | 4.251 | 3.444 | 133.5 | 35.59 |
| 600.0 | 8.0771 | 3345 | 3022 | 6.011 | 3.597 | 2.538 | 130.0 | 36.90 |
| 650.0 | 9.0459 | 3516 | 3154 | 6.201 | 3.256 | 2.035 | 131.5 | 38.47 |
| 700.0 | 9.9295 | 3673 | 3276 | 6.367 | 3.057 | 1.714 | 134.8 | 40.13 |
| 750.0 | 10.754 | 3823 | 3393 | 6.517 | 2.933 | 1.490 | 138.8 | 41.83 |
| 800.0 | 11.536 | 3967 | 3506 | 6.655 | 2.854 | 1.325 | 143.1 | 43.52 |
| 850.0 | 12.286 | 4109 | 3617 | 6.784 | 2.803 | 1.198 | 147.4 | 45.19 |
| 900.0 | 13.010 | 4248 | 3728 | 6.905 | 2.772 | 1.096 | 151.7 | 46.85 |
| 950.0 | 13.714 | 4386 | 3837 | 7.020 | 2.753 | 1.014 | 155.8 | 48.48 |
| 1000.0 | 14.401 | 4523 | 3947 | 7.130 | 2.744 | 0.944 | 160.0 | 50.08 |

Table 1 (cont.)

$$p = 42.5\,\text{MPa} = 425\ \text{bar}$$

| $t$ | $v$ | $h$ | $u$ | $s$ | $c_p$ | $\alpha_p$ | $\lambda$ | $\eta$ |
|---|---|---|---|---|---|---|---|---|
| °C | dm³/kg | kJ/kg | kJ/kg | kJ/kg K | kJ/kg K | $10^{-3}$/K | mW/K m | $\mu$Pa s |
| 0.0 | 0.9800 | 42.0 | 0.4 | −0.0001 | 4.044 | 0.073 | 584.2 | 1707 |
| 5.0 | 0.9805 | 62.2 | 20.6 | 0.073 | 4.042 | 0.131 | 592.1 | 1463 |
| 10.0 | 0.9813 | 82.5 | 40.8 | 0.145 | 4.050 | 0.181 | 600.5 | 1271 |
| 15.0 | 0.9823 | 102.7 | 61.0 | 0.216 | 4.060 | 0.225 | 609.2 | 1116 |
| 20.0 | 0.9835 | 123.1 | 81.3 | 0.286 | 4.070 | 0.266 | 617.9 | 989.8 |
| 25.0 | 0.9849 | 143.4 | 101.6 | 0.355 | 4.078 | 0.302 | 626.4 | 885.0 |
| 30.0 | 0.9865 | 163.9 | 121.9 | 0.423 | 4.085 | 0.336 | 634.6 | 797.0 |
| 35.0 | 0.9882 | 184.3 | 142.3 | 0.490 | 4.089 | 0.368 | 642.5 | 722.5 |
| 40.0 | 0.9901 | 204.7 | 162.7 | 0.556 | 4.092 | 0.398 | 649.9 | 658.6 |
| 45.0 | 0.9921 | 225.2 | 183.0 | 0.620 | 4.094 | 0.426 | 656.8 | 603.5 |
| 50.0 | 0.9943 | 245.7 | 203.4 | 0.684 | 4.096 | 0.454 | 663.2 | 555.6 |
| 60.0 | 0.9991 | 286.7 | 244.2 | 0.809 | 4.099 | 0.506 | 674.5 | 476.8 |
| 70.0 | 1.0044 | 327.7 | 285.0 | 0.931 | 4.104 | 0.554 | 683.8 | 415.1 |
| 80.0 | 1.0102 | 368.7 | 325.8 | 1.048 | 4.111 | 0.601 | 691.4 | 365.9 |
| 90.0 | 1.0166 | 409.9 | 366.7 | 1.163 | 4.120 | 0.645 | 697.4 | 326.0 |
| 100.0 | 1.0234 | 451.1 | 407.6 | 1.275 | 4.130 | 0.689 | 702.0 | 293.2 |
| 125.0 | 1.0425 | 554.8 | 510.4 | 1.544 | 4.160 | 0.795 | 708.7 | 233.0 |
| 150.0 | 1.0649 | 659.2 | 613.9 | 1.799 | 4.197 | 0.901 | 709.8 | 192.8 |
| 175.0 | 1.0906 | 764.7 | 718.3 | 2.041 | 4.242 | 1.015 | 705.9 | 164.7 |
| 200.0 | 1.1204 | 871.4 | 823.8 | 2.273 | 4.303 | 1.142 | 697.3 | 144.0 |
| 225.0 | 1.1549 | 980.0 | 930.9 | 2.496 | 4.386 | 1.291 | 683.7 | 128.3 |
| 250.0 | 1.1954 | 1090 | 1040 | 2.713 | 4.500 | 1.473 | 664.7 | 115.8 |
| 275.0 | 1.2437 | 1205 | 1152 | 2.927 | 4.655 | 1.703 | 639.7 | 105.5 |
| 300.0 | 1.3025 | 1324 | 1268 | 3.139 | 4.870 | 2.012 | 608.2 | 96.64 |
| 325.0 | 1.3768 | 1449 | 1391 | 3.353 | 5.177 | 2.450 | 569.9 | 88.55 |
| 350.0 | 1.4752 | 1584 | 1521 | 3.574 | 5.643 | 3.128 | 526.5 | 80.72 |
| 375.0 | 1.6166 | 1734 | 1665 | 3.810 | 6.433 | 4.319 | 480.0 | 72.58 |
| 400.0 | 1.8499 | 1912 | 1833 | 4.079 | 7.997 | 6.800 | 425.2 | 63.33 |
| 450.0 | 3.2455 | 2440 | 2302 | 4.834 | 11.30 | 12.99 | 253.2 | 42.02 |
| 500.0 | 5.0912 | 2860 | 2644 | 5.397 | 6.208 | 6.004 | 163.4 | 36.27 |
| 550.0 | 6.4174 | 3119 | 2846 | 5.723 | 4.446 | 3.649 | 140.5 | 36.28 |
| 600.0 | 7.4902 | 3321 | 3002 | 5.961 | 3.713 | 2.650 | 135.2 | 37.41 |
| 650.0 | 8.4257 | 3496 | 3138 | 6.156 | 3.334 | 2.105 | 135.8 | 38.89 |
| 700.0 | 9.2752 | 3657 | 3263 | 6.326 | 3.114 | 1.761 | 138.7 | 40.49 |
| 750.0 | 10.066 | 3809 | 3381 | 6.478 | 2.977 | 1.524 | 142.3 | 42.14 |
| 800.0 | 10.813 | 3955 | 3496 | 6.618 | 2.889 | 1.350 | 146.3 | 43.80 |
| 850.0 | 11.529 | 4098 | 3608 | 6.748 | 2.832 | 1.217 | 150.3 | 45.46 |
| 900.0 | 12.218 | 4239 | 3719 | 6.870 | 2.795 | 1.111 | 154.3 | 47.09 |
| 950.0 | 12.888 | 4378 | 3830 | 6.987 | 2.773 | 1.025 | 158.3 | 48.70 |
| 1000.0 | 13.541 | 4516 | 3941 | 7.097 | 2.761 | 0.954 | 162.1 | 50.29 |

Table 1 (cont.)

$$p = 45.0\,\text{MPa} = 450\ \text{bar}$$

| $t$ | $v$ | $h$ | $u$ | $s$ | $c_p$ | $\alpha_p$ | $\lambda$ | $\eta$ |
|---|---|---|---|---|---|---|---|---|
| °C | dm³/kg | kJ/kg | kJ/kg | kJ/kg K | kJ/kg K | $10^{-3}$/K | mW/K m | $\mu$Pa s |
| 0.0 | 0.9789 | 44.4 | 0.4 | −0.0003 | 4.036 | 0.081 | 585.5 | 1704 |
| 5.0 | 0.9795 | 64.6 | 20.5 | 0.073 | 4.035 | 0.137 | 593.4 | 1461 |
| 10.0 | 0.9802 | 84.8 | 40.7 | 0.145 | 4.043 | 0.186 | 601.7 | 1269 |
| 15.0 | 0.9813 | 105.0 | 60.9 | 0.216 | 4.054 | 0.229 | 610.4 | 1115 |
| 20.0 | 0.9825 | 125.3 | 81.1 | 0.286 | 4.065 | 0.269 | 619.0 | 989.4 |
| 25.0 | 0.9839 | 145.7 | 101.4 | 0.354 | 4.073 | 0.305 | 627.5 | 884.9 |
| 30.0 | 0.9855 | 166.1 | 121.7 | 0.422 | 4.080 | 0.338 | 635.8 | 797.2 |
| 35.0 | 0.9872 | 186.5 | 142.1 | 0.489 | 4.084 | 0.369 | 643.6 | 722.8 |
| 40.0 | 0.9891 | 206.9 | 162.4 | 0.555 | 4.087 | 0.399 | 651.0 | 659.1 |
| 45.0 | 0.9912 | 227.3 | 182.7 | 0.619 | 4.089 | 0.427 | 657.9 | 604.1 |
| 50.0 | 0.9934 | 247.8 | 203.1 | 0.683 | 4.091 | 0.454 | 664.4 | 556.2 |
| 60.0 | 0.9981 | 288.7 | 243.8 | 0.808 | 4.095 | 0.505 | 675.7 | 477.5 |
| 70.0 | 1.0034 | 329.7 | 284.5 | 0.929 | 4.100 | 0.553 | 685.0 | 415.8 |
| 80.0 | 1.0092 | 370.7 | 325.3 | 1.047 | 4.107 | 0.599 | 692.6 | 366.5 |
| 90.0 | 1.0155 | 411.8 | 366.1 | 1.162 | 4.115 | 0.643 | 698.7 | 326.6 |
| 100.0 | 1.0223 | 453.0 | 407.0 | 1.274 | 4.125 | 0.686 | 703.3 | 293.9 |
| 125.0 | 1.0413 | 556.5 | 509.7 | 1.542 | 4.155 | 0.790 | 710.2 | 233.6 |
| 150.0 | 1.0635 | 660.9 | 613.0 | 1.796 | 4.191 | 0.895 | 711.4 | 193.4 |
| 175.0 | 1.0890 | 766.2 | 717.1 | 2.038 | 4.235 | 1.007 | 707.7 | 165.2 |
| 200.0 | 1.1185 | 872.7 | 822.4 | 2.269 | 4.294 | 1.131 | 699.2 | 144.6 |
| 225.0 | 1.1526 | 981.0 | 929.2 | 2.492 | 4.374 | 1.276 | 685.9 | 128.8 |
| 250.0 | 1.1925 | 1091 | 1038 | 2.709 | 4.484 | 1.452 | 667.2 | 116.4 |
| 275.0 | 1.2399 | 1205 | 1149 | 2.922 | 4.633 | 1.674 | 642.7 | 106.1 |
| 300.0 | 1.2974 | 1323 | 1265 | 3.133 | 4.837 | 1.966 | 611.7 | 97.29 |
| 325.0 | 1.3693 | 1448 | 1386 | 3.345 | 5.123 | 2.375 | 574.2 | 89.28 |
| 350.0 | 1.4634 | 1581 | 1515 | 3.563 | 5.547 | 2.989 | 531.7 | 81.60 |
| 375.0 | 1.5954 | 1727 | 1655 | 3.793 | 6.233 | 4.016 | 486.5 | 73.74 |
| 400.0 | 1.8025 | 1897 | 1816 | 4.050 | 7.483 | 5.977 | 435.0 | 65.07 |
| 450.0 | 2.9123 | 2377 | 2246 | 4.736 | 10.94 | 12.18 | 276.6 | 45.06 |
| 500.0 | 4.6324 | 2814 | 2605 | 5.322 | 6.601 | 6.438 | 176.1 | 37.56 |
| 550.0 | 5.9281 | 3087 | 2820 | 5.665 | 4.644 | 3.852 | 147.9 | 37.03 |
| 600.0 | 6.9713 | 3296 | 2982 | 5.912 | 3.829 | 2.761 | 140.6 | 37.95 |
| 650.0 | 7.8764 | 3476 | 3122 | 6.112 | 3.412 | 2.173 | 140.3 | 39.32 |
| 700.0 | 8.6951 | 3640 | 3249 | 6.286 | 3.170 | 1.808 | 142.6 | 40.86 |
| 750.0 | 9.4548 | 3795 | 3369 | 6.440 | 3.020 | 1.557 | 145.9 | 42.47 |
| 800.0 | 10.172 | 3943 | 3485 | 6.582 | 2.923 | 1.375 | 149.6 | 44.10 |
| 850.0 | 10.856 | 4088 | 3599 | 6.714 | 2.860 | 1.236 | 153.3 | 45.73 |
| 900.0 | 11.515 | 4229 | 3711 | 6.837 | 2.819 | 1.126 | 157.1 | 47.34 |
| 950.0 | 12.154 | 4370 | 3823 | 6.954 | 2.793 | 1.037 | 160.7 | 48.93 |
| 1000.0 | 12.777 | 4509 | 3934 | 7.066 | 2.778 | 0.963 | 164.4 | 50.51 |

Table 1 (cont.)

$p = 47.5\,\text{MPa} = 475\ \text{bar}$

| $t$ | $v$ | $h$ | $u$ | $s$ | $c_p$ | $\alpha_p$ | $\lambda$ | $\eta$ |
|---|---|---|---|---|---|---|---|---|
| °C | dm³/kg | kJ/kg | kJ/kg | kJ/kg K | kJ/kg K | $10^{-3}$/K | mW/K m | $\mu$Pa s |
| 0.0 | 0.9778 | 46.8 | 0.4 | −0.0005 | 4.028 | 0.088 | 586.8 | 1700 |
| 5.0 | 0.9784 | 67.0 | 20.5 | 0.073 | 4.028 | 0.143 | 594.6 | 1459 |
| 10.0 | 0.9792 | 87.1 | 40.6 | 0.144 | 4.037 | 0.190 | 602.9 | 1268 |
| 15.0 | 0.9802 | 107.3 | 60.8 | 0.215 | 4.048 | 0.233 | 611.5 | 1114 |
| 20.0 | 0.9815 | 127.6 | 81.0 | 0.285 | 4.059 | 0.272 | 620.1 | 989.1 |
| 25.0 | 0.9829 | 147.9 | 101.2 | 0.354 | 4.068 | 0.307 | 628.6 | 884.9 |
| 30.0 | 0.9845 | 168.3 | 121.5 | 0.421 | 4.075 | 0.340 | 636.9 | 797.4 |
| 35.0 | 0.9862 | 188.7 | 141.8 | 0.488 | 4.079 | 0.370 | 644.7 | 723.1 |
| 40.0 | 0.9882 | 209.1 | 162.1 | 0.554 | 4.083 | 0.399 | 652.1 | 659.5 |
| 45.0 | 0.9902 | 229.5 | 182.5 | 0.618 | 4.085 | 0.427 | 659.1 | 604.6 |
| 50.0 | 0.9924 | 249.9 | 202.8 | 0.682 | 4.087 | 0.454 | 665.5 | 556.8 |
| 60.0 | 0.9971 | 290.8 | 243.4 | 0.807 | 4.090 | 0.504 | 676.9 | 478.1 |
| 70.0 | 1.0024 | 331.7 | 284.1 | 0.928 | 4.095 | 0.552 | 686.2 | 416.5 |
| 80.0 | 1.0082 | 372.7 | 324.8 | 1.045 | 4.102 | 0.597 | 693.9 | 367.2 |
| 90.0 | 1.0145 | 413.8 | 365.6 | 1.160 | 4.111 | 0.641 | 699.9 | 327.3 |
| 100.0 | 1.0212 | 454.9 | 406.4 | 1.272 | 4.121 | 0.683 | 704.7 | 294.5 |
| 125.0 | 1.0401 | 558.3 | 508.9 | 1.540 | 4.150 | 0.786 | 711.6 | 234.3 |
| 150.0 | 1.0621 | 662.5 | 612.1 | 1.794 | 4.185 | 0.889 | 713.0 | 194.0 |
| 175.0 | 1.0875 | 767.6 | 716.0 | 2.035 | 4.228 | 0.999 | 709.5 | 165.8 |
| 200.0 | 1.1166 | 874.0 | 821.0 | 2.266 | 4.285 | 1.121 | 701.2 | 145.1 |
| 225.0 | 1.1504 | 982.1 | 927.4 | 2.489 | 4.363 | 1.262 | 688.0 | 129.4 |
| 250.0 | 1.1897 | 1092 | 1035 | 2.705 | 4.469 | 1.433 | 669.6 | 116.9 |
| 275.0 | 1.2362 | 1205 | 1147 | 2.917 | 4.611 | 1.646 | 645.5 | 106.7 |
| 300.0 | 1.2924 | 1323 | 1262 | 3.126 | 4.805 | 1.924 | 615.2 | 97.91 |
| 325.0 | 1.3622 | 1446 | 1381 | 3.337 | 5.073 | 2.306 | 578.4 | 89.99 |
| 350.0 | 1.4525 | 1578 | 1509 | 3.552 | 5.461 | 2.866 | 536.8 | 82.45 |
| 375.0 | 1.5764 | 1721 | 1646 | 3.778 | 6.065 | 3.765 | 492.6 | 74.83 |
| 400.0 | 1.7634 | 1884 | 1801 | 4.025 | 7.094 | 5.365 | 443.6 | 66.62 |
| 450.0 | 2.6674 | 2326 | 2199 | 4.656 | 10.30 | 10.95 | 297.4 | 47.91 |
| 500.0 | 4.2348 | 2768 | 2567 | 5.248 | 6.953 | 6.802 | 189.3 | 38.97 |
| 550.0 | 5.4952 | 3055 | 2794 | 5.608 | 4.841 | 4.049 | 155.7 | 37.84 |
| 600.0 | 6.5100 | 3272 | 2962 | 5.864 | 3.946 | 2.870 | 146.3 | 38.53 |
| 650.0 | 7.3869 | 3456 | 3105 | 6.070 | 3.490 | 2.241 | 145.0 | 39.78 |
| 700.0 | 8.1776 | 3624 | 3235 | 6.247 | 3.227 | 1.853 | 146.7 | 41.25 |
| 750.0 | 8.9093 | 3781 | 3357 | 6.404 | 3.063 | 1.590 | 149.6 | 42.81 |
| 800.0 | 9.5984 | 3931 | 3475 | 6.548 | 2.958 | 1.399 | 152.9 | 44.40 |
| 850.0 | 10.255 | 4077 | 3590 | 6.681 | 2.888 | 1.254 | 156.4 | 46.00 |
| 900.0 | 10.887 | 4220 | 3703 | 6.805 | 2.842 | 1.141 | 159.8 | 47.59 |
| 950.0 | 11.498 | 4362 | 3815 | 6.923 | 2.813 | 1.049 | 163.3 | 49.17 |
| 1000.0 | 12.094 | 4502 | 3927 | 7.036 | 2.795 | 0.972 | 166.6 | 50.73 |

Table 1 (cont.)

$$p = 50.0\,\text{MPa} = 500\ \text{bar}$$

| $t$ | $v$ | $h$ | $u$ | $s$ | $c_p$ | $\alpha_p$ | $\lambda$ | $\eta$ |
|---|---|---|---|---|---|---|---|---|
| °C | dm³/kg | kJ/kg | kJ/kg | kJ/kg K | kJ/kg K | $10^{-3}$/K | mW/K m | $\mu$Pa s |
| 0.0 | 0.9767 | 49.2 | 0.4 | −0.0008 | 4.020 | 0.096 | 588.1 | 1697 |
| 5.0 | 0.9773 | 69.3 | 20.4 | 0.072 | 4.021 | 0.149 | 595.8 | 1456 |
| 10.0 | 0.9782 | 89.4 | 40.5 | 0.144 | 4.031 | 0.195 | 604.1 | 1267 |
| 15.0 | 0.9792 | 109.6 | 60.6 | 0.215 | 4.043 | 0.237 | 612.7 | 1114 |
| 20.0 | 0.9805 | 129.9 | 80.8 | 0.284 | 4.054 | 0.275 | 621.3 | 988.8 |
| 25.0 | 0.9819 | 150.1 | 101.1 | 0.353 | 4.063 | 0.309 | 629.8 | 884.9 |
| 30.0 | 0.9835 | 170.5 | 121.3 | 0.420 | 4.070 | 0.341 | 638.0 | 797.6 |
| 35.0 | 0.9853 | 190.8 | 141.6 | 0.487 | 4.075 | 0.372 | 645.9 | 723.5 |
| 40.0 | 0.9872 | 211.2 | 161.9 | 0.553 | 4.078 | 0.400 | 653.3 | 660.0 |
| 45.0 | 0.9892 | 231.6 | 182.2 | 0.617 | 4.081 | 0.428 | 660.2 | 605.1 |
| 50.0 | 0.9914 | 252.0 | 202.5 | 0.681 | 4.083 | 0.454 | 666.7 | 557.4 |
| 60.0 | 0.9962 | 292.9 | 243.1 | 0.805 | 4.086 | 0.503 | 678.0 | 478.8 |
| 70.0 | 1.0014 | 333.8 | 283.7 | 0.926 | 4.091 | 0.550 | 687.4 | 417.1 |
| 80.0 | 1.0072 | 374.7 | 324.3 | 1.044 | 4.098 | 0.595 | 695.1 | 367.9 |
| 90.0 | 1.0134 | 415.7 | 365.1 | 1.159 | 4.107 | 0.638 | 701.2 | 328.0 |
| 100.0 | 1.0201 | 456.8 | 405.8 | 1.270 | 4.117 | 0.680 | 706.0 | 295.2 |
| 125.0 | 1.0389 | 560.1 | 508.2 | 1.538 | 4.146 | 0.782 | 713.1 | 234.9 |
| 150.0 | 1.0608 | 664.2 | 611.1 | 1.792 | 4.180 | 0.884 | 714.6 | 194.6 |
| 175.0 | 1.0859 | 769.2 | 714.9 | 2.033 | 4.221 | 0.991 | 711.2 | 166.4 |
| 200.0 | 1.1148 | 875.3 | 819.6 | 2.263 | 4.277 | 1.111 | 703.2 | 145.7 |
| 225.0 | 1.1481 | 983.2 | 925.7 | 2.485 | 4.352 | 1.249 | 690.2 | 129.9 |
| 250.0 | 1.1869 | 1093 | 1033 | 2.701 | 4.454 | 1.414 | 672.1 | 117.5 |
| 275.0 | 1.2326 | 1206 | 1144 | 2.912 | 4.591 | 1.619 | 648.4 | 107.3 |
| 300.0 | 1.2876 | 1323 | 1258 | 3.120 | 4.775 | 1.884 | 618.5 | 98.53 |
| 325.0 | 1.3554 | 1445 | 1377 | 3.329 | 5.027 | 2.242 | 582.5 | 90.68 |
| 350.0 | 1.4422 | 1575 | 1503 | 3.542 | 5.384 | 2.756 | 541.7 | 83.27 |
| 375.0 | 1.5593 | 1716 | 1638 | 3.763 | 5.921 | 3.551 | 498.4 | 75.85 |
| 400.0 | 1.7301 | 1874 | 1787 | 4.002 | 6.789 | 4.890 | 451.5 | 68.02 |
| 450.0 | 2.4858 | 2284 | 2160 | 4.589 | 9.595 | 9.699 | 315.8 | 50.50 |
| 500.0 | 3.8919 | 2724 | 2529 | 5.178 | 7.239 | 7.064 | 202.8 | 40.49 |
| 550.0 | 5.1112 | 3023 | 2767 | 5.554 | 5.033 | 4.235 | 163.7 | 38.71 |
| 600.0 | 6.0981 | 3247 | 2942 | 5.818 | 4.062 | 2.976 | 152.1 | 39.14 |
| 650.0 | 6.9486 | 3437 | 3089 | 6.030 | 3.567 | 2.308 | 149.8 | 40.26 |
| 700.0 | 7.7134 | 3607 | 3222 | 6.210 | 3.283 | 1.898 | 150.9 | 41.65 |
| 750.0 | 8.4196 | 3767 | 3346 | 6.369 | 3.107 | 1.622 | 153.4 | 43.16 |
| 800.0 | 9.0834 | 3919 | 3465 | 6.515 | 2.992 | 1.423 | 156.4 | 44.72 |
| 850.0 | 9.7151 | 4067 | 3581 | 6.649 | 2.916 | 1.273 | 159.5 | 46.29 |
| 900.0 | 10.322 | 4211 | 3695 | 6.775 | 2.866 | 1.155 | 162.7 | 47.86 |
| 950.0 | 10.908 | 4353 | 3808 | 6.894 | 2.833 | 1.060 | 165.9 | 49.41 |
| 1000.0 | 11.479 | 4495 | 3921 | 7.007 | 2.812 | 0.982 | 169.0 | 50.95 |

Table 1 (cont.)

$p = 55.0\,\text{MPa} = 550\ \text{bar}$

| $t$ | $v$ | $h$ | $u$ | $s$ | $c_p$ | $\alpha_p$ | $\lambda$ | $\eta$ |
|---|---|---|---|---|---|---|---|---|
| °C | dm³/kg | kJ/kg | kJ/kg | kJ/kg K | kJ/kg K | $10^{-3}$/K | mW/K m | $\mu$Pa s |
| 0.0 | 0.9746 | 53.9 | 0.3 | −0.001 | 4.006 | 0.111 | 590.7 | 1690 |
| 5.0 | 0.9752 | 74.0 | 20.3 | 0.071 | 4.008 | 0.160 | 598.3 | 1452 |
| 10.0 | 0.9761 | 94.0 | 40.3 | 0.143 | 4.019 | 0.205 | 606.4 | 1264 |
| 15.0 | 0.9772 | 114.2 | 60.4 | 0.213 | 4.032 | 0.244 | 614.9 | 1112 |
| 20.0 | 0.9785 | 134.4 | 80.5 | 0.283 | 4.044 | 0.280 | 623.5 | 988.3 |
| 25.0 | 0.9800 | 154.6 | 100.7 | 0.351 | 4.053 | 0.314 | 632.0 | 884.9 |
| 30.0 | 0.9816 | 174.9 | 120.9 | 0.419 | 4.061 | 0.345 | 640.2 | 798.1 |
| 35.0 | 0.9834 | 195.2 | 141.1 | 0.485 | 4.066 | 0.374 | 648.1 | 724.2 |
| 40.0 | 0.9853 | 215.5 | 161.3 | 0.551 | 4.069 | 0.402 | 655.5 | 660.9 |
| 45.0 | 0.9873 | 235.9 | 181.6 | 0.615 | 4.072 | 0.428 | 662.5 | 606.2 |
| 50.0 | 0.9895 | 256.3 | 201.8 | 0.679 | 4.074 | 0.454 | 668.9 | 558.6 |
| 60.0 | 0.9942 | 297.0 | 242.3 | 0.803 | 4.078 | 0.502 | 680.4 | 480.1 |
| 70.0 | 0.9995 | 337.8 | 282.9 | 0.924 | 4.083 | 0.548 | 689.8 | 418.5 |
| 80.0 | 1.0052 | 378.7 | 323.4 | 1.041 | 4.090 | 0.592 | 697.6 | 369.3 |
| 90.0 | 1.0114 | 419.6 | 364.0 | 1.155 | 4.098 | 0.634 | 703.7 | 329.4 |
| 100.0 | 1.0180 | 460.7 | 404.7 | 1.267 | 4.108 | 0.675 | 708.6 | 296.5 |
| 125.0 | 1.0366 | 563.7 | 506.7 | 1.534 | 4.136 | 0.774 | 715.9 | 236.1 |
| 150.0 | 1.0581 | 667.5 | 609.3 | 1.787 | 4.169 | 0.873 | 717.7 | 195.8 |
| 175.0 | 1.0829 | 772.2 | 712.6 | 2.027 | 4.208 | 0.977 | 714.7 | 167.5 |
| 200.0 | 1.1112 | 878.0 | 816.9 | 2.257 | 4.260 | 1.091 | 707.0 | 146.7 |
| 225.0 | 1.1438 | 985.4 | 922.5 | 2.478 | 4.331 | 1.223 | 694.5 | 131.0 |
| 250.0 | 1.1815 | 1094 | 1029 | 2.692 | 4.426 | 1.378 | 676.9 | 118.6 |
| 275.0 | 1.2257 | 1206 | 1139 | 2.902 | 4.552 | 1.569 | 653.9 | 108.4 |
| 300.0 | 1.2784 | 1322 | 1252 | 3.108 | 4.720 | 1.810 | 625.0 | 99.73 |
| 325.0 | 1.3427 | 1443 | 1369 | 3.314 | 4.944 | 2.129 | 590.4 | 92.01 |
| 350.0 | 1.4234 | 1570 | 1492 | 3.523 | 5.251 | 2.568 | 551.0 | 84.81 |
| 375.0 | 1.5292 | 1706 | 1622 | 3.737 | 5.687 | 3.208 | 509.3 | 77.75 |
| 400.0 | 1.6759 | 1856 | 1764 | 3.964 | 6.335 | 4.197 | 465.4 | 70.49 |
| 450.0 | 2.2407 | 2223 | 2099 | 4.488 | 8.399 | 7.644 | 346.6 | 54.89 |
| 500.0 | 3.3475 | 2642 | 2458 | 5.049 | 7.552 | 7.209 | 229.9 | 43.72 |
| 550.0 | 4.4662 | 2960 | 2715 | 5.449 | 5.386 | 4.553 | 180.6 | 40.59 |
| 600.0 | 5.3966 | 3199 | 2902 | 5.730 | 4.289 | 3.173 | 164.3 | 40.44 |
| 650.0 | 6.1983 | 3398 | 3057 | 5.952 | 3.721 | 2.434 | 159.8 | 41.27 |
| 700.0 | 6.9165 | 3575 | 3194 | 6.139 | 3.394 | 1.985 | 159.6 | 42.49 |
| 750.0 | 7.5774 | 3739 | 3322 | 6.304 | 3.192 | 1.684 | 161.2 | 43.89 |
| 800.0 | 8.1966 | 3895 | 3444 | 6.452 | 3.060 | 1.469 | 163.4 | 45.36 |
| 850.0 | 8.7843 | 4046 | 3563 | 6.590 | 2.972 | 1.308 | 165.9 | 46.87 |
| 900.0 | 9.3474 | 4193 | 3679 | 6.718 | 2.912 | 1.182 | 168.5 | 48.39 |
| 950.0 | 9.8908 | 4337 | 3793 | 6.838 | 2.872 | 1.082 | 171.1 | 49.91 |
| 1000.0 | 10.418 | 4480 | 3907 | 6.953 | 2.846 | 0.999 | 173.7 | 51.41 |

Table 1 (cont.)

$$p = 60.0\,\text{MPa} = 600\ \text{bar}$$

| $t$ | $v$ | $h$ | $u$ | $s$ | $c_p$ | $\alpha_p$ | $\lambda$ | $\eta$ |
|---|---|---|---|---|---|---|---|---|
| °C | dm³/kg | kJ/kg | kJ/kg | kJ/kg K | kJ/kg K | $10^{-3}$/K | mW/K m | $\mu$Pa s |
| 0.0 | 0.9725 | 58.7 | 0.3 | −0.002 | 3.992 | 0.125 | 593.3 | 1684 |
| 5.0 | 0.9732 | 78.6 | 20.2 | 0.071 | 3.996 | 0.172 | 600.7 | 1449 |
| 10.0 | 0.9741 | 98.6 | 40.2 | 0.142 | 4.008 | 0.214 | 608.8 | 1262 |
| 15.0 | 0.9753 | 118.7 | 60.2 | 0.212 | 4.021 | 0.251 | 617.2 | 1111 |
| 20.0 | 0.9766 | 138.8 | 80.2 | 0.281 | 4.034 | 0.286 | 625.8 | 987.9 |
| 25.0 | 0.9781 | 159.0 | 100.3 | 0.350 | 4.044 | 0.318 | 634.2 | 885.1 |
| 30.0 | 0.9797 | 179.3 | 120.5 | 0.417 | 4.052 | 0.348 | 642.4 | 798.6 |
| 35.0 | 0.9815 | 199.5 | 140.7 | 0.483 | 4.057 | 0.376 | 650.3 | 725.1 |
| 40.0 | 0.9834 | 219.8 | 160.8 | 0.549 | 4.061 | 0.403 | 657.8 | 662.0 |
| 45.0 | 0.9854 | 240.2 | 181.0 | 0.613 | 4.064 | 0.429 | 664.7 | 607.4 |
| 50.0 | 0.9876 | 260.5 | 201.2 | 0.676 | 4.066 | 0.454 | 671.2 | 559.9 |
| 60.0 | 0.9923 | 301.2 | 241.6 | 0.800 | 4.070 | 0.501 | 682.7 | 481.4 |
| 70.0 | 0.9975 | 341.9 | 282.0 | 0.921 | 4.075 | 0.546 | 692.2 | 419.9 |
| 80.0 | 1.0032 | 382.7 | 322.5 | 1.038 | 4.082 | 0.588 | 700.0 | 370.7 |
| 90.0 | 1.0093 | 423.5 | 363.0 | 1.152 | 4.090 | 0.629 | 706.3 | 330.7 |
| 100.0 | 1.0159 | 464.5 | 403.5 | 1.263 | 4.100 | 0.669 | 711.2 | 297.9 |
| 125.0 | 1.0343 | 567.3 | 505.2 | 1.530 | 4.127 | 0.766 | 718.8 | 237.4 |
| 150.0 | 1.0556 | 670.8 | 607.5 | 1.782 | 4.158 | 0.862 | 720.9 | 197.0 |
| 175.0 | 1.0799 | 775.2 | 710.4 | 2.022 | 4.196 | 0.963 | 718.2 | 168.6 |
| 200.0 | 1.1077 | 880.7 | 814.3 | 2.251 | 4.245 | 1.073 | 710.9 | 147.8 |
| 225.0 | 1.1395 | 987.6 | 919.3 | 2.471 | 4.311 | 1.198 | 698.7 | 132.0 |
| 250.0 | 1.1763 | 1096 | 1025 | 2.684 | 4.400 | 1.345 | 681.6 | 119.6 |
| 275.0 | 1.2191 | 1207 | 1134 | 2.892 | 4.517 | 1.523 | 659.2 | 109.5 |
| 300.0 | 1.2698 | 1322 | 1246 | 3.097 | 4.670 | 1.745 | 631.3 | 100.9 |
| 325.0 | 1.3310 | 1441 | 1361 | 3.300 | 4.871 | 2.030 | 597.8 | 93.29 |
| 350.0 | 1.4066 | 1566 | 1482 | 3.505 | 5.138 | 2.412 | 559.7 | 86.26 |
| 375.0 | 1.5034 | 1699 | 1609 | 3.714 | 5.502 | 2.943 | 519.4 | 79.47 |
| 400.0 | 1.6328 | 1843 | 1745 | 3.931 | 6.011 | 3.711 | 477.6 | 72.65 |
| 450.0 | 2.0839 | 2180 | 2054 | 4.413 | 7.547 | 6.233 | 371.2 | 58.45 |
| 500.0 | 2.9547 | 2571 | 2394 | 4.937 | 7.534 | 6.901 | 255.8 | 47.03 |
| 550.0 | 3.9548 | 2901 | 2663 | 5.351 | 5.673 | 4.765 | 198.1 | 42.64 |
| 600.0 | 4.8263 | 3152 | 2862 | 5.647 | 4.500 | 3.343 | 177.0 | 41.85 |
| 650.0 | 5.5823 | 3359 | 3025 | 5.879 | 3.869 | 2.549 | 170.1 | 42.36 |
| 700.0 | 6.2590 | 3543 | 3167 | 6.072 | 3.503 | 2.064 | 168.5 | 43.38 |
| 750.0 | 6.8804 | 3712 | 3299 | 6.242 | 3.275 | 1.742 | 169.1 | 44.65 |
| 800.0 | 7.4614 | 3872 | 3424 | 6.394 | 3.126 | 1.512 | 170.6 | 46.04 |
| 850.0 | 8.0115 | 4025 | 3545 | 6.534 | 3.026 | 1.341 | 172.4 | 47.49 |
| 900.0 | 8.5377 | 4175 | 3663 | 6.664 | 2.957 | 1.208 | 174.4 | 48.95 |
| 950.0 | 9.0446 | 4321 | 3779 | 6.787 | 2.910 | 1.102 | 176.5 | 50.42 |
| 1000.0 | 9.5359 | 4466 | 3894 | 6.903 | 2.879 | 1.016 | 178.6 | 51.89 |

Table 1 (cont.) $p = 65.0\,\text{MPa} = 650\ \text{bar}$

| $t$ | $v$ | $h$ | $u$ | $s$ | $c_p$ | $\alpha_p$ | $\lambda$ | $\eta$ |
|------|--------|-------|-------|--------|--------|-----------|--------|--------|
| °C | dm³/kg | kJ/kg | kJ/kg | kJ/kg K | kJ/kg K | $10^{-3}$/K | mW/K m | $\mu$Pa s |
| 0.0 | 0.9704 | 63.3 | 0.3 | −0.002 | 3.979 | 0.138 | 595.8 | 1679 |
| 5.0 | 0.9712 | 83.2 | 20.1 | 0.070 | 3.984 | 0.182 | 603.1 | 1445 |
| 10.0 | 0.9721 | 103.2 | 40.0 | 0.141 | 3.997 | 0.222 | 611.0 | 1260 |
| 15.0 | 0.9733 | 123.2 | 59.9 | 0.211 | 4.011 | 0.258 | 619.4 | 1110 |
| 20.0 | 0.9747 | 143.3 | 79.9 | 0.280 | 4.024 | 0.292 | 628.0 | 987.7 |
| 25.0 | 0.9762 | 163.4 | 100.0 | 0.348 | 4.035 | 0.323 | 636.4 | 885.4 |
| 30.0 | 0.9778 | 183.6 | 120.1 | 0.415 | 4.043 | 0.351 | 644.6 | 799.2 |
| 35.0 | 0.9796 | 203.9 | 140.2 | 0.482 | 4.049 | 0.379 | 652.5 | 725.9 |
| 40.0 | 0.9815 | 224.1 | 160.3 | 0.547 | 4.053 | 0.405 | 660.0 | 663.0 |
| 45.0 | 0.9836 | 244.4 | 180.5 | 0.611 | 4.056 | 0.430 | 667.0 | 608.6 |
| 50.0 | 0.9857 | 264.7 | 200.6 | 0.674 | 4.058 | 0.454 | 673.5 | 561.1 |
| 60.0 | 0.9905 | 305.3 | 240.9 | 0.798 | 4.062 | 0.500 | 685.0 | 482.8 |
| 70.0 | 0.9956 | 345.9 | 281.2 | 0.918 | 4.067 | 0.543 | 694.6 | 421.3 |
| 80.0 | 1.0013 | 386.6 | 321.5 | 1.035 | 4.074 | 0.585 | 702.5 | 372.1 |
| 90.0 | 1.0073 | 427.4 | 361.9 | 1.149 | 4.082 | 0.625 | 708.8 | 332.1 |
| 100.0 | 1.0139 | 468.3 | 402.4 | 1.260 | 4.091 | 0.664 | 713.8 | 299.2 |
| 125.0 | 1.0321 | 570.9 | 503.8 | 1.526 | 4.118 | 0.758 | 721.6 | 238.6 |
| 150.0 | 1.0530 | 674.2 | 605.8 | 1.778 | 4.148 | 0.852 | 724.0 | 198.1 |
| 175.0 | 1.0770 | 778.3 | 708.3 | 2.017 | 4.184 | 0.949 | 721.7 | 169.7 |
| 200.0 | 1.1043 | 883.5 | 811.7 | 2.245 | 4.230 | 1.055 | 714.7 | 148.9 |
| 225.0 | 1.1355 | 990.0 | 916.1 | 2.464 | 4.292 | 1.175 | 702.9 | 133.1 |
| 250.0 | 1.1713 | 1098 | 1022 | 2.677 | 4.375 | 1.314 | 686.3 | 120.7 |
| 275.0 | 1.2129 | 1208 | 1130 | 2.883 | 4.484 | 1.481 | 664.5 | 110.6 |
| 300.0 | 1.2617 | 1322 | 1240 | 3.086 | 4.625 | 1.685 | 637.4 | 102.0 |
| 325.0 | 1.3201 | 1440 | 1354 | 3.287 | 4.807 | 1.944 | 604.9 | 94.51 |
| 350.0 | 1.3914 | 1563 | 1473 | 3.489 | 5.042 | 2.279 | 568.0 | 87.63 |
| 375.0 | 1.4808 | 1693 | 1596 | 3.693 | 5.352 | 2.730 | 528.8 | 81.06 |
| 400.0 | 1.5971 | 1831 | 1728 | 3.903 | 5.765 | 3.350 | 488.5 | 74.57 |
| 450.0 | 1.9741 | 2148 | 2019 | 4.355 | 6.945 | 5.264 | 391.4 | 61.42 |
| 500.0 | 2.6717 | 2513 | 2340 | 4.844 | 7.304 | 6.342 | 279.8 | 50.20 |
| 550.0 | 3.5491 | 2845 | 2614 | 5.260 | 5.870 | 4.852 | 215.5 | 44.81 |
| 600.0 | 4.3586 | 3106 | 2823 | 5.569 | 4.688 | 3.475 | 189.9 | 43.36 |
| 650.0 | 5.0706 | 3322 | 2993 | 5.809 | 4.008 | 2.648 | 180.6 | 43.50 |
| 700.0 | 5.7095 | 3512 | 3141 | 6.010 | 3.607 | 2.136 | 177.6 | 44.32 |
| 750.0 | 6.2957 | 3685 | 3276 | 6.184 | 3.355 | 1.795 | 177.2 | 45.46 |
| 800.0 | 6.8431 | 3849 | 3404 | 6.339 | 3.191 | 1.553 | 177.9 | 46.75 |
| 850.0 | 7.3607 | 4005 | 3527 | 6.482 | 3.079 | 1.372 | 179.1 | 48.12 |
| 900.0 | 7.8549 | 4157 | 3647 | 6.614 | 3.002 | 1.233 | 180.5 | 49.53 |
| 950.0 | 8.3304 | 4306 | 3764 | 6.738 | 2.948 | 1.122 | 182.0 | 50.95 |
| 1000.0 | 8.7907 | 4452 | 3881 | 6.856 | 2.911 | 1.032 | 183.5 | 52.38 |

Table 1 (cont.)

$$p = 70.0\,\text{MPa} = 700\ \text{bar}$$

| $t$ | $v$ | $h$ | $u$ | $s$ | $c_p$ | $\alpha_p$ | $\lambda$ | $\eta$ |
|---|---|---|---|---|---|---|---|---|
| °C | dm³/kg | kJ/kg | kJ/kg | kJ/kg K | kJ/kg K | $10^{-3}$/K | mW/K m | $\mu$Pa s |
| 0.0 | 0.9683 | 68.0 | 0.2 | −0.003 | 3.967 | 0.151 | 598.3 | 1674 |
| 5.0 | 0.9692 | 87.8 | 20.0 | 0.069 | 3.973 | 0.193 | 605.4 | 1442 |
| 10.0 | 0.9702 | 107.7 | 39.8 | 0.140 | 3.987 | 0.231 | 613.3 | 1258 |
| 15.0 | 0.9714 | 127.7 | 59.7 | 0.210 | 4.002 | 0.265 | 621.7 | 1109 |
| 20.0 | 0.9728 | 147.7 | 79.7 | 0.279 | 4.015 | 0.297 | 630.2 | 987.6 |
| 25.0 | 0.9743 | 167.9 | 99.7 | 0.347 | 4.026 | 0.327 | 638.6 | 885.8 |
| 30.0 | 0.9759 | 188.0 | 119.7 | 0.414 | 4.035 | 0.355 | 646.8 | 799.9 |
| 35.0 | 0.9777 | 208.2 | 139.8 | 0.480 | 4.041 | 0.381 | 654.7 | 726.9 |
| 40.0 | 0.9797 | 228.4 | 159.8 | 0.545 | 4.045 | 0.406 | 662.2 | 664.1 |
| 45.0 | 0.9817 | 248.6 | 179.9 | 0.609 | 4.048 | 0.431 | 669.2 | 609.8 |
| 50.0 | 0.9839 | 268.9 | 200.0 | 0.672 | 4.050 | 0.454 | 675.7 | 562.4 |
| 60.0 | 0.9886 | 309.4 | 240.2 | 0.796 | 4.054 | 0.499 | 687.3 | 484.2 |
| 70.0 | 0.9938 | 350.0 | 280.4 | 0.915 | 4.059 | 0.541 | 696.9 | 422.7 |
| 80.0 | 0.9994 | 390.6 | 320.6 | 1.032 | 4.066 | 0.582 | 704.9 | 373.4 |
| 90.0 | 1.0054 | 431.3 | 360.9 | 1.146 | 4.074 | 0.621 | 711.3 | 333.4 |
| 100.0 | 1.0118 | 472.1 | 401.3 | 1.257 | 4.083 | 0.659 | 716.4 | 300.5 |
| 125.0 | 1.0298 | 574.5 | 502.4 | 1.522 | 4.109 | 0.751 | 724.5 | 239.9 |
| 150.0 | 1.0505 | 677.6 | 604.0 | 1.773 | 4.138 | 0.842 | 727.2 | 199.3 |
| 175.0 | 1.0741 | 781.4 | 706.2 | 2.012 | 4.172 | 0.936 | 725.1 | 170.8 |
| 200.0 | 1.1010 | 886.2 | 809.2 | 2.239 | 4.216 | 1.038 | 718.5 | 149.9 |
| 225.0 | 1.1315 | 992.3 | 913.1 | 2.458 | 4.274 | 1.153 | 707.1 | 134.1 |
| 250.0 | 1.1665 | 1100 | 1018 | 2.669 | 4.352 | 1.285 | 690.8 | 121.7 |
| 275.0 | 1.2068 | 1210 | 1125 | 2.874 | 4.454 | 1.442 | 669.6 | 111.6 |
| 300.0 | 1.2540 | 1323 | 1235 | 3.076 | 4.584 | 1.631 | 643.2 | 103.1 |
| 325.0 | 1.3099 | 1439 | 1347 | 3.275 | 4.749 | 1.867 | 611.7 | 95.68 |
| 350.0 | 1.3774 | 1560 | 1464 | 3.473 | 4.959 | 2.165 | 575.9 | 88.93 |
| 375.0 | 1.4608 | 1688 | 1585 | 3.673 | 5.227 | 2.554 | 537.6 | 82.55 |
| 400.0 | 1.5667 | 1822 | 1713 | 3.877 | 5.571 | 3.068 | 498.6 | 76.33 |
| 450.0 | 1.8917 | 2123 | 1991 | 4.308 | 6.505 | 4.572 | 408.3 | 63.98 |
| 500.0 | 2.4645 | 2466 | 2294 | 4.767 | 6.992 | 5.721 | 301.7 | 53.13 |
| 550.0 | 3.2267 | 2794 | 2569 | 5.178 | 5.974 | 4.816 | 232.5 | 47.03 |
| 600.0 | 3.9725 | 3063 | 2785 | 5.496 | 4.846 | 3.563 | 202.7 | 44.93 |
| 650.0 | 4.6416 | 3286 | 2961 | 5.744 | 4.135 | 2.729 | 191.0 | 44.70 |
| 700.0 | 5.2452 | 3481 | 3114 | 5.950 | 3.705 | 2.199 | 186.7 | 45.30 |
| 750.0 | 5.7996 | 3659 | 3253 | 6.129 | 3.433 | 1.842 | 185.3 | 46.29 |
| 800.0 | 6.3170 | 3826 | 3384 | 6.288 | 3.253 | 1.589 | 185.2 | 47.47 |
| 850.0 | 6.8058 | 3986 | 3509 | 6.433 | 3.130 | 1.401 | 185.7 | 48.77 |
| 900.0 | 7.2720 | 4140 | 3631 | 6.567 | 3.045 | 1.256 | 186.5 | 50.12 |
| 950.0 | 7.7202 | 4290 | 3750 | 6.693 | 2.985 | 1.140 | 187.4 | 51.50 |
| 1000.0 | 8.1534 | 4439 | 3868 | 6.812 | 2.943 | 1.047 | 188.4 | 52.88 |

Table 1 (cont.)

$p = 75.0\,\text{MPa} = 750\ \text{bar}$

| $t$ | $v$ | $h$ | $u$ | $s$ | $c_p$ | $\alpha_p$ | $\lambda$ | $\eta$ |
|------|------|------|------|------|------|------|------|------|
| °C | dm³/kg | kJ/kg | kJ/kg | kJ/kg K | kJ/kg K | $10^{-3}$/K | mW/K m | μPa s |
| 0.0 | 0.9663 | 72.6 | 0.1 | −0.004 | 3.956 | 0.164 | 600.7 | 1669 |
| 5.0 | 0.9672 | 92.4 | 19.9 | 0.068 | 3.963 | 0.203 | 607.7 | 1440 |
| 10.0 | 0.9683 | 112.3 | 39.6 | 0.139 | 3.977 | 0.239 | 615.6 | 1257 |
| 15.0 | 0.9695 | 132.2 | 59.5 | 0.208 | 3.993 | 0.272 | 623.9 | 1109 |
| 20.0 | 0.9709 | 152.2 | 79.4 | 0.277 | 4.007 | 0.302 | 632.3 | 987.7 |
| 25.0 | 0.9724 | 172.2 | 99.3 | 0.345 | 4.018 | 0.331 | 640.8 | 886.3 |
| 30.0 | 0.9741 | 192.4 | 119.3 | 0.412 | 4.027 | 0.358 | 649.0 | 800.7 |
| 35.0 | 0.9759 | 212.5 | 139.3 | 0.478 | 4.033 | 0.383 | 656.9 | 727.9 |
| 40.0 | 0.9779 | 232.7 | 159.3 | 0.543 | 4.037 | 0.408 | 664.4 | 665.3 |
| 45.0 | 0.9799 | 252.9 | 179.4 | 0.607 | 4.040 | 0.432 | 671.4 | 611.1 |
| 50.0 | 0.9821 | 273.1 | 199.4 | 0.670 | 4.043 | 0.454 | 678.0 | 563.8 |
| 60.0 | 0.9868 | 313.5 | 239.5 | 0.793 | 4.047 | 0.498 | 689.6 | 485.6 |
| 70.0 | 0.9919 | 354.0 | 279.6 | 0.913 | 4.052 | 0.539 | 699.3 | 424.1 |
| 80.0 | 0.9975 | 394.6 | 319.8 | 1.029 | 4.058 | 0.579 | 707.3 | 374.8 |
| 90.0 | 1.0034 | 435.2 | 359.9 | 1.143 | 4.066 | 0.617 | 713.8 | 334.8 |
| 100.0 | 1.0098 | 475.9 | 400.2 | 1.253 | 4.075 | 0.654 | 719.0 | 301.8 |
| 125.0 | 1.0277 | 578.1 | 501.0 | 1.518 | 4.101 | 0.744 | 727.3 | 241.1 |
| 150.0 | 1.0481 | 681.0 | 602.4 | 1.769 | 4.128 | 0.832 | 730.3 | 200.4 |
| 175.0 | 1.0714 | 784.6 | 704.2 | 2.007 | 4.161 | 0.924 | 728.5 | 171.8 |
| 200.0 | 1.0977 | 889.1 | 806.7 | 2.234 | 4.202 | 1.022 | 722.2 | 150.9 |
| 225.0 | 1.1277 | 994.8 | 910.2 | 2.451 | 4.257 | 1.132 | 711.2 | 135.1 |
| 250.0 | 1.1618 | 1102 | 1014 | 2.662 | 4.330 | 1.258 | 695.3 | 122.7 |
| 275.0 | 1.2011 | 1211 | 1121 | 2.866 | 4.425 | 1.406 | 674.6 | 112.6 |
| 300.0 | 1.2467 | 1323 | 1230 | 3.066 | 4.545 | 1.582 | 648.9 | 104.2 |
| 325.0 | 1.3004 | 1438 | 1341 | 3.263 | 4.697 | 1.798 | 618.3 | 96.82 |
| 350.0 | 1.3645 | 1558 | 1456 | 3.459 | 4.885 | 2.066 | 583.4 | 90.17 |
| 375.0 | 1.4427 | 1683 | 1575 | 3.655 | 5.120 | 2.406 | 546.0 | 83.95 |
| 400.0 | 1.5402 | 1815 | 1699 | 3.854 | 5.413 | 2.842 | 507.9 | 77.95 |
| 450.0 | 1.8269 | 2104 | 1966 | 4.268 | 6.172 | 4.059 | 422.9 | 66.23 |
| 500.0 | 2.3088 | 2428 | 2255 | 4.702 | 6.674 | 5.139 | 321.4 | 55.80 |
| 550.0 | 2.9697 | 2749 | 2527 | 5.105 | 5.995 | 4.683 | 249.0 | 49.24 |
| 600.0 | 3.6520 | 3023 | 2749 | 5.427 | 4.969 | 3.604 | 215.3 | 46.56 |
| 650.0 | 4.2792 | 3252 | 2931 | 5.683 | 4.249 | 2.790 | 201.4 | 45.95 |
| 700.0 | 4.8495 | 3452 | 3088 | 5.894 | 3.796 | 2.251 | 195.6 | 46.31 |
| 750.0 | 5.3746 | 3634 | 3231 | 6.077 | 3.506 | 1.883 | 193.3 | 47.14 |
| 800.0 | 5.8648 | 3804 | 3364 | 6.239 | 3.312 | 1.622 | 192.4 | 48.22 |
| 850.0 | 6.3278 | 3966 | 3492 | 6.387 | 3.179 | 1.427 | 192.3 | 49.44 |
| 900.0 | 6.7692 | 4123 | 3615 | 6.523 | 3.086 | 1.277 | 192.5 | 50.72 |
| 950.0 | 7.1931 | 4275 | 3736 | 6.650 | 3.021 | 1.157 | 192.9 | 52.05 |
| 1000.0 | 7.6026 | 4425 | 3855 | 6.770 | 2.974 | 1.061 | 193.4 | 53.40 |

Table 1 (cont.) $p = 80.0\,\text{MPa} = 800\ \text{bar}$

| $t$ | $v$ | $h$ | $u$ | $s$ | $c_p$ | $\alpha_p$ | $\lambda$ | $\eta$ |
|---|---|---|---|---|---|---|---|---|
| °C | dm³/kg | kJ/kg | kJ/kg | kJ/kg K | kJ/kg K | $10^{-3}$/K | mW/K m | $\mu$Pa s |
| 0.0 | 0.9643 | 77.2 | 0.08 | −0.005 | 3.945 | 0.175 | 603.1 | 1665 |
| 5.0 | 0.9653 | 97.0 | 19.7 | 0.067 | 3.953 | 0.213 | 610.0 | 1437 |
| 10.0 | 0.9664 | 116.8 | 39.5 | 0.137 | 3.968 | 0.247 | 617.8 | 1256 |
| 15.0 | 0.9676 | 136.6 | 59.2 | 0.207 | 3.984 | 0.278 | 626.0 | 1108 |
| 20.0 | 0.9691 | 156.6 | 79.1 | 0.276 | 3.998 | 0.308 | 634.5 | 987.9 |
| 25.0 | 0.9706 | 176.6 | 99.0 | 0.343 | 4.010 | 0.335 | 642.9 | 886.8 |
| 30.0 | 0.9723 | 196.7 | 118.9 | 0.410 | 4.019 | 0.361 | 651.2 | 801.6 |
| 35.0 | 0.9741 | 216.8 | 138.9 | 0.476 | 4.025 | 0.386 | 659.1 | 728.9 |
| 40.0 | 0.9761 | 236.9 | 158.9 | 0.541 | 4.030 | 0.409 | 666.6 | 666.5 |
| 45.0 | 0.9781 | 257.1 | 178.8 | 0.605 | 4.033 | 0.432 | 673.6 | 612.3 |
| 50.0 | 0.9803 | 277.3 | 198.8 | 0.668 | 4.035 | 0.454 | 680.2 | 565.1 |
| 60.0 | 0.9850 | 317.6 | 238.8 | 0.791 | 4.040 | 0.497 | 691.8 | 487.0 |
| 70.0 | 0.9901 | 358.1 | 278.9 | 0.910 | 4.045 | 0.537 | 701.6 | 425.5 |
| 80.0 | 0.9956 | 398.5 | 318.9 | 1.026 | 4.051 | 0.576 | 709.7 | 376.2 |
| 90.0 | 1.0015 | 439.1 | 359.0 | 1.140 | 4.059 | 0.613 | 716.2 | 336.2 |
| 100.0 | 1.0079 | 479.7 | 399.1 | 1.250 | 4.068 | 0.650 | 721.5 | 303.2 |
| 125.0 | 1.0255 | 581.7 | 499.7 | 1.515 | 4.093 | 0.737 | 730.1 | 242.3 |
| 150.0 | 1.0457 | 684.4 | 600.7 | 1.765 | 4.119 | 0.823 | 733.4 | 201.5 |
| 175.0 | 1.0686 | 787.7 | 702.2 | 2.002 | 4.150 | 0.912 | 732.0 | 172.9 |
| 200.0 | 1.0946 | 891.9 | 804.3 | 2.228 | 4.189 | 1.007 | 725.9 | 152.0 |
| 225.0 | 1.1239 | 997.2 | 907.3 | 2.445 | 4.241 | 1.113 | 715.3 | 136.1 |
| 250.0 | 1.1573 | 1104 | 1011 | 2.654 | 4.309 | 1.232 | 699.8 | 123.7 |
| 275.0 | 1.1955 | 1212 | 1117 | 2.857 | 4.398 | 1.372 | 679.6 | 113.6 |
| 300.0 | 1.2397 | 1324 | 1225 | 3.056 | 4.510 | 1.537 | 654.5 | 105.2 |
| 325.0 | 1.2914 | 1438 | 1335 | 3.251 | 4.649 | 1.735 | 624.6 | 97.92 |
| 350.0 | 1.3526 | 1556 | 1448 | 3.445 | 4.820 | 1.978 | 590.6 | 91.37 |
| 375.0 | 1.4263 | 1679 | 1565 | 3.639 | 5.028 | 2.279 | 554.1 | 85.28 |
| 400.0 | 1.5168 | 1808 | 1687 | 3.833 | 5.280 | 2.655 | 516.7 | 79.46 |
| 450.0 | 1.7740 | 2087 | 1946 | 4.233 | 5.912 | 3.663 | 435.7 | 68.26 |
| 500.0 | 2.1882 | 2397 | 2222 | 4.647 | 6.383 | 4.632 | 339.1 | 58.22 |
| 550.0 | 2.7634 | 2710 | 2489 | 5.039 | 5.955 | 4.484 | 264.6 | 51.39 |
| 600.0 | 3.3846 | 2985 | 2714 | 5.364 | 5.057 | 3.602 | 227.5 | 48.20 |
| 650.0 | 3.9711 | 3219 | 2901 | 5.625 | 4.347 | 2.830 | 211.5 | 47.22 |
| 700.0 | 4.5098 | 3424 | 3063 | 5.841 | 3.880 | 2.292 | 204.4 | 47.34 |
| 750.0 | 5.0076 | 3610 | 3209 | 6.027 | 3.574 | 1.918 | 201.1 | 48.01 |
| 800.0 | 5.4729 | 3783 | 3345 | 6.193 | 3.369 | 1.650 | 199.6 | 48.98 |
| 850.0 | 5.9126 | 3947 | 3474 | 6.343 | 3.227 | 1.450 | 198.8 | 50.12 |
| 900.0 | 6.3316 | 4106 | 3600 | 6.481 | 3.126 | 1.295 | 198.4 | 51.34 |
| 950.0 | 6.7339 | 4261 | 3722 | 6.610 | 3.055 | 1.173 | 198.2 | 52.62 |
| 1000.0 | 7.1222 | 4412 | 3842 | 6.731 | 3.004 | 1.073 | 198.2 | 53.92 |

Table 1 (cont.)

$$p = 85.0\,\text{MPa} = 850\ \text{bar}$$

| $t$ | $v$ | $h$ | $u$ | $s$ | $c_p$ | $\alpha_p$ | $\lambda$ | $\eta$ |
|---|---|---|---|---|---|---|---|---|
| °C | dm³/kg | kJ/kg | kJ/kg | kJ/kg K | kJ/kg K | $10^{-3}$/K | mW/K m | $\mu$Pa s |
| 0.0 | 0.9624 | 81.8 | −0.00003 | −0.006 | 3.936 | 0.187 | 605.5 | 1661 |
| 5.0 | 0.9633 | 101.5 | 19.6 | 0.066 | 3.943 | 0.222 | 612.2 | 1435 |
| 10.0 | 0.9645 | 121.2 | 39.3 | 0.136 | 3.959 | 0.254 | 620.0 | 1255 |
| 15.0 | 0.9658 | 141.1 | 59.0 | 0.206 | 3.975 | 0.284 | 628.2 | 1108 |
| 20.0 | 0.9672 | 161.0 | 78.8 | 0.274 | 3.990 | 0.313 | 636.6 | 988.1 |
| 25.0 | 0.9688 | 181.0 | 98.6 | 0.342 | 4.002 | 0.339 | 645.1 | 887.5 |
| 30.0 | 0.9705 | 201.0 | 118.5 | 0.408 | 4.011 | 0.364 | 653.3 | 802.5 |
| 35.0 | 0.9723 | 221.1 | 138.4 | 0.474 | 4.018 | 0.388 | 661.2 | 730.1 |
| 40.0 | 0.9743 | 241.2 | 158.4 | 0.539 | 4.022 | 0.411 | 668.7 | 667.7 |
| 45.0 | 0.9763 | 261.3 | 178.3 | 0.603 | 4.026 | 0.433 | 675.8 | 613.7 |
| 50.0 | 0.9785 | 281.4 | 198.3 | 0.665 | 4.028 | 0.455 | 682.4 | 566.5 |
| 60.0 | 0.9832 | 321.7 | 238.2 | 0.788 | 4.032 | 0.496 | 694.1 | 488.4 |
| 70.0 | 0.9883 | 362.1 | 278.1 | 0.907 | 4.037 | 0.535 | 703.9 | 426.9 |
| 80.0 | 0.9938 | 402.5 | 318.0 | 1.024 | 4.044 | 0.573 | 712.1 | 377.6 |
| 90.0 | 0.9997 | 443.0 | 358.0 | 1.137 | 4.052 | 0.610 | 718.7 | 337.5 |
| 100.0 | 1.0059 | 483.5 | 398.0 | 1.247 | 4.060 | 0.645 | 724.0 | 304.5 |
| 125.0 | 1.0234 | 585.3 | 498.4 | 1.511 | 4.085 | 0.731 | 732.9 | 243.5 |
| 150.0 | 1.0434 | 687.8 | 599.1 | 1.760 | 4.110 | 0.814 | 736.4 | 202.7 |
| 175.0 | 1.0660 | 790.9 | 700.3 | 1.997 | 4.139 | 0.901 | 735.4 | 174.0 |
| 200.0 | 1.0915 | 894.8 | 802.0 | 2.223 | 4.176 | 0.993 | 729.7 | 153.0 |
| 225.0 | 1.1203 | 999.8 | 904.6 | 2.439 | 4.225 | 1.094 | 719.3 | 137.1 |
| 250.0 | 1.1529 | 1106 | 1008 | 2.647 | 4.290 | 1.208 | 704.2 | 124.7 |
| 275.0 | 1.1902 | 1214 | 1113 | 2.849 | 4.373 | 1.340 | 684.4 | 114.6 |
| 300.0 | 1.2331 | 1324 | 1220 | 3.047 | 4.477 | 1.495 | 659.9 | 106.2 |
| 325.0 | 1.2829 | 1438 | 1329 | 3.240 | 4.606 | 1.679 | 630.7 | 98.98 |
| 350.0 | 1.3415 | 1555 | 1441 | 3.432 | 4.761 | 1.900 | 597.5 | 92.52 |
| 375.0 | 1.4113 | 1676 | 1556 | 3.623 | 4.947 | 2.170 | 561.7 | 86.55 |
| 400.0 | 1.4959 | 1803 | 1675 | 3.814 | 5.168 | 2.498 | 525.1 | 80.88 |
| 450.0 | 1.7297 | 2074 | 1927 | 4.203 | 5.702 | 3.350 | 447.2 | 70.10 |
| 500.0 | 2.0923 | 2371 | 2194 | 4.600 | 6.128 | 4.203 | 355.1 | 60.43 |
| 550.0 | 2.5964 | 2675 | 2454 | 4.981 | 5.874 | 4.252 | 279.5 | 53.47 |
| 600.0 | 3.1605 | 2950 | 2681 | 5.305 | 5.112 | 3.562 | 239.3 | 49.85 |
| 650.0 | 3.7075 | 3188 | 2873 | 5.570 | 4.430 | 2.850 | 221.2 | 48.52 |
| 700.0 | 4.2161 | 3397 | 3038 | 5.791 | 3.955 | 2.322 | 212.9 | 48.40 |
| 750.0 | 4.6884 | 3586 | 3187 | 5.980 | 3.638 | 1.947 | 208.8 | 48.90 |
| 800.0 | 5.1308 | 3762 | 3326 | 6.149 | 3.422 | 1.674 | 206.5 | 49.76 |
| 850.0 | 5.5491 | 3929 | 3457 | 6.301 | 3.271 | 1.470 | 205.1 | 50.80 |
| 900.0 | 5.9478 | 4090 | 3584 | 6.441 | 3.165 | 1.312 | 204.2 | 51.96 |
| 950.0 | 6.3305 | 4246 | 3708 | 6.571 | 3.088 | 1.187 | 203.5 | 53.19 |
| 1000.0 | 6.6998 | 4399 | 3830 | 6.694 | 3.033 | 1.085 | 203.0 | 54.44 |

Table 1 (cont.)

$$p = 90.0\,\text{MPa} = 900\ \text{bar}$$

| $t$ | $v$ | $h$ | $u$ | $s$ | $c_p$ | $\alpha_p$ | $\lambda$ | $\eta$ |
|------|------|------|------|------|------|------|------|------|
| °C | dm³/kg | kJ/kg | kJ/kg | kJ/kg K | kJ/kg K | $10^{-3}$/K | mW/K m | µPa s |
| 0.0 | 0.9604 | 86.4 | −0.08 | −0.007 | 3.926 | 0.198 | 607.7 | 1658 |
| 5.0 | 0.9615 | 106.0 | 19.5 | 0.065 | 3.934 | 0.231 | 614.5 | 1433 |
| 10.0 | 0.9626 | 125.7 | 39.1 | 0.135 | 3.950 | 0.262 | 622.1 | 1254 |
| 15.0 | 0.9640 | 145.5 | 58.7 | 0.204 | 3.967 | 0.290 | 630.3 | 1108 |
| 20.0 | 0.9654 | 165.4 | 78.5 | 0.273 | 3.983 | 0.317 | 638.7 | 988.6 |
| 25.0 | 0.9670 | 185.3 | 98.3 | 0.340 | 3.995 | 0.343 | 647.2 | 888.3 |
| 30.0 | 0.9688 | 205.3 | 118.1 | 0.407 | 4.004 | 0.367 | 655.4 | 803.5 |
| 35.0 | 0.9706 | 225.4 | 138.0 | 0.472 | 4.011 | 0.390 | 663.3 | 731.2 |
| 40.0 | 0.9725 | 245.4 | 157.9 | 0.537 | 4.015 | 0.412 | 670.9 | 669.0 |
| 45.0 | 0.9746 | 265.5 | 177.8 | 0.600 | 4.019 | 0.434 | 678.0 | 615.0 |
| 50.0 | 0.9768 | 285.6 | 197.7 | 0.663 | 4.021 | 0.455 | 684.6 | 567.9 |
| 60.0 | 0.9814 | 325.8 | 237.5 | 0.786 | 4.026 | 0.495 | 696.3 | 489.9 |
| 70.0 | 0.9865 | 366.1 | 277.3 | 0.905 | 4.031 | 0.534 | 706.2 | 428.4 |
| 80.0 | 0.9919 | 406.5 | 317.2 | 1.021 | 4.037 | 0.570 | 714.4 | 379.0 |
| 90.0 | 0.9978 | 446.9 | 357.1 | 1.134 | 4.045 | 0.606 | 721.1 | 338.9 |
| 100.0 | 1.0040 | 487.4 | 397.0 | 1.244 | 4.053 | 0.641 | 726.6 | 305.8 |
| 125.0 | 1.0213 | 589.0 | 497.1 | 1.507 | 4.077 | 0.724 | 735.6 | 244.7 |
| 150.0 | 1.0410 | 691.2 | 597.5 | 1.756 | 4.101 | 0.806 | 739.5 | 203.8 |
| 175.0 | 1.0633 | 794.1 | 698.4 | 1.992 | 4.129 | 0.890 | 738.7 | 175.0 |
| 200.0 | 1.0884 | 897.7 | 799.7 | 2.217 | 4.164 | 0.979 | 733.4 | 154.0 |
| 225.0 | 1.1167 | 1002 | 901.8 | 2.433 | 4.210 | 1.076 | 723.3 | 138.1 |
| 250.0 | 1.1487 | 1108 | 1004 | 2.640 | 4.271 | 1.185 | 708.6 | 125.6 |
| 275.0 | 1.1851 | 1216 | 1109 | 2.841 | 4.349 | 1.311 | 689.2 | 115.6 |
| 300.0 | 1.2267 | 1325 | 1215 | 3.038 | 4.447 | 1.456 | 665.1 | 107.2 |
| 325.0 | 1.2748 | 1438 | 1323 | 3.230 | 4.565 | 1.627 | 636.6 | 100.0 |
| 350.0 | 1.3311 | 1554 | 1434 | 3.420 | 4.708 | 1.830 | 604.2 | 93.62 |
| 375.0 | 1.3975 | 1674 | 1548 | 3.608 | 4.876 | 2.073 | 569.1 | 87.75 |
| 400.0 | 1.4770 | 1798 | 1665 | 3.796 | 5.071 | 2.363 | 533.0 | 82.22 |
| 450.0 | 1.6916 | 2063 | 1910 | 4.175 | 5.529 | 3.095 | 457.5 | 71.80 |
| 500.0 | 2.0140 | 2350 | 2169 | 4.559 | 5.908 | 3.843 | 369.6 | 62.46 |
| 550.0 | 2.4594 | 2645 | 2423 | 4.928 | 5.770 | 4.008 | 293.5 | 55.45 |
| 600.0 | 2.9715 | 2918 | 2651 | 5.251 | 5.138 | 3.491 | 250.6 | 51.49 |
| 650.0 | 3.4808 | 3158 | 2845 | 5.519 | 4.495 | 2.851 | 230.6 | 49.82 |
| 700.0 | 3.9607 | 3371 | 3014 | 5.743 | 4.022 | 2.341 | 221.1 | 49.46 |
| 750.0 | 4.4091 | 3563 | 3166 | 5.936 | 3.696 | 1.968 | 216.1 | 49.81 |
| 800.0 | 4.8302 | 3742 | 3307 | 6.107 | 3.472 | 1.694 | 213.1 | 50.54 |
| 850.0 | 5.2288 | 3911 | 3441 | 6.261 | 3.314 | 1.487 | 211.2 | 51.50 |
| 900.0 | 5.6089 | 4074 | 3569 | 6.403 | 3.201 | 1.327 | 209.8 | 52.59 |
| 950.0 | 5.9738 | 4232 | 3694 | 6.535 | 3.120 | 1.199 | 208.6 | 53.76 |
| 1000.0 | 6.3259 | 4386 | 3817 | 6.658 | 3.061 | 1.095 | 207.6 | 54.97 |

Table 1 (cont.)

$$p = 95.0\,\mathrm{MPa} = 950\ \mathrm{bar}$$

| $t$ | $v$ | $h$ | $u$ | $s$ | $c_p$ | $\alpha_p$ | $\lambda$ | $\eta$ |
|------|------|------|------|------|------|------|------|------|
| °C | dm³/kg | kJ/kg | kJ/kg | kJ/kg K | kJ/kg K | $10^{-3}$/K | mW/K m | $\mu$Pa s |
| 0.0 | 0.9585 | 90.9 | −0.2 | −0.008 | 3.917 | 0.208 | 610.0 | 1655 |
| 5.0 | 0.9596 | 110.5 | 19.3 | 0.064 | 3.926 | 0.239 | 616.6 | 1432 |
| 10.0 | 0.9608 | 130.2 | 38.9 | 0.134 | 3.942 | 0.268 | 624.2 | 1253 |
| 15.0 | 0.9622 | 149.9 | 58.5 | 0.203 | 3.960 | 0.296 | 632.4 | 1108 |
| 20.0 | 0.9637 | 169.8 | 78.2 | 0.271 | 3.975 | 0.322 | 640.8 | 989.1 |
| 25.0 | 0.9653 | 189.7 | 98.0 | 0.338 | 3.988 | 0.346 | 649.3 | 889.1 |
| 30.0 | 0.9670 | 209.6 | 117.8 | 0.405 | 3.997 | 0.370 | 657.5 | 804.6 |
| 35.0 | 0.9689 | 229.6 | 137.6 | 0.470 | 4.004 | 0.392 | 665.5 | 732.5 |
| 40.0 | 0.9708 | 249.7 | 157.4 | 0.535 | 4.009 | 0.414 | 673.0 | 670.3 |
| 45.0 | 0.9729 | 269.7 | 177.3 | 0.598 | 4.012 | 0.435 | 680.1 | 616.4 |
| 50.0 | 0.9750 | 289.8 | 197.1 | 0.661 | 4.014 | 0.455 | 686.8 | 569.3 |
| 60.0 | 0.9797 | 329.9 | 236.9 | 0.783 | 4.019 | 0.494 | 698.6 | 491.3 |
| 70.0 | 0.9847 | 370.2 | 276.6 | 0.902 | 4.024 | 0.532 | 708.5 | 429.8 |
| 80.0 | 0.9901 | 410.4 | 316.4 | 1.018 | 4.030 | 0.568 | 716.8 | 380.5 |
| 90.0 | 0.9960 | 450.8 | 356.1 | 1.131 | 4.038 | 0.603 | 723.6 | 340.3 |
| 100.0 | 1.0022 | 491.2 | 396.0 | 1.240 | 4.046 | 0.637 | 729.1 | 307.1 |
| 125.0 | 1.0193 | 592.6 | 495.8 | 1.503 | 4.069 | 0.718 | 738.4 | 245.9 |
| 150.0 | 1.0388 | 694.6 | 596.0 | 1.752 | 4.093 | 0.798 | 742.6 | 204.9 |
| 175.0 | 1.0608 | 797.3 | 696.5 | 1.988 | 4.119 | 0.879 | 742.1 | 176.1 |
| 200.0 | 1.0855 | 900.6 | 797.5 | 2.212 | 4.152 | 0.965 | 737.1 | 155.0 |
| 225.0 | 1.1133 | 1004 | 899.2 | 2.427 | 4.196 | 1.059 | 727.3 | 139.0 |
| 250.0 | 1.1446 | 1110 | 1001 | 2.634 | 4.253 | 1.164 | 712.9 | 126.6 |
| 275.0 | 1.1801 | 1217 | 1105 | 2.834 | 4.327 | 1.283 | 693.9 | 116.5 |
| 300.0 | 1.2206 | 1327 | 1211 | 3.029 | 4.418 | 1.420 | 670.3 | 108.2 |
| 325.0 | 1.2672 | 1438 | 1318 | 3.220 | 4.529 | 1.580 | 642.4 | 101.0 |
| 350.0 | 1.3212 | 1553 | 1428 | 3.408 | 4.659 | 1.767 | 610.6 | 94.69 |
| 375.0 | 1.3846 | 1671 | 1540 | 3.594 | 4.812 | 1.987 | 576.1 | 88.91 |
| 400.0 | 1.4597 | 1794 | 1655 | 3.779 | 4.986 | 2.246 | 540.6 | 83.50 |
| 450.0 | 1.6585 | 2053 | 1895 | 4.150 | 5.385 | 2.883 | 467.1 | 73.38 |
| 500.0 | 1.9487 | 2331 | 2146 | 4.522 | 5.720 | 3.540 | 382.7 | 64.33 |
| 550.0 | 2.3458 | 2618 | 2395 | 4.882 | 5.656 | 3.769 | 306.8 | 57.33 |
| 600.0 | 2.8112 | 2889 | 2622 | 5.202 | 5.140 | 3.397 | 261.5 | 53.10 |
| 650.0 | 3.2850 | 3131 | 2819 | 5.471 | 4.545 | 2.834 | 239.5 | 51.13 |
| 700.0 | 3.7376 | 3346 | 2991 | 5.698 | 4.080 | 2.350 | 228.8 | 50.54 |
| 750.0 | 4.1633 | 3541 | 3146 | 5.894 | 3.749 | 1.983 | 223.1 | 50.72 |
| 800.0 | 4.5645 | 3723 | 3289 | 6.067 | 3.518 | 1.709 | 219.5 | 51.33 |
| 850.0 | 4.9448 | 3894 | 3424 | 6.223 | 3.354 | 1.501 | 217.0 | 52.20 |
| 900.0 | 5.3078 | 4059 | 3554 | 6.366 | 3.236 | 1.339 | 215.1 | 53.22 |
| 950.0 | 5.6564 | 4218 | 3681 | 6.500 | 3.150 | 1.210 | 213.5 | 54.34 |
| 1000.0 | 5.9927 | 4374 | 3805 | 6.625 | 3.087 | 1.104 | 212.1 | 55.51 |

Table 1 (cont.)

$$p = 100.0 \, \text{MPa} = 1000 \, \text{bar}$$

| $t$ | $v$ | $h$ | $u$ | $s$ | $c_p$ | $\alpha_p$ | $\lambda$ | $\eta$ |
|---|---|---|---|---|---|---|---|---|
| °C | dm³/kg | kJ/kg | kJ/kg | kJ/kg K | kJ/kg K | $10^{-3}$/K | mW/K m | µPa s |
| 0.0 | 0.9567 | 95.4 | −0.3 | −0.009 | 3.909 | 0.217 | 612.2 | 1652 |
| 5.0 | 0.9578 | 115.0 | 19.2 | 0.062 | 3.918 | 0.247 | 618.7 | 1430 |
| 10.0 | 0.9590 | 134.6 | 38.7 | 0.132 | 3.935 | 0.275 | 626.3 | 1253 |
| 15.0 | 0.9604 | 154.3 | 58.3 | 0.201 | 3.952 | 0.302 | 634.5 | 1108 |
| 20.0 | 0.9619 | 174.1 | 77.9 | 0.270 | 3.968 | 0.326 | 642.9 | 989.7 |
| 25.0 | 0.9635 | 194.0 | 97.6 | 0.337 | 3.981 | 0.350 | 651.3 | 890.1 |
| 30.0 | 0.9653 | 213.9 | 117.4 | 0.403 | 3.990 | 0.373 | 659.6 | 805.7 |
| 35.0 | 0.9671 | 233.9 | 137.2 | 0.468 | 3.997 | 0.394 | 667.6 | 733.7 |
| 40.0 | 0.9691 | 253.9 | 157.0 | 0.533 | 4.002 | 0.415 | 675.1 | 671.7 |
| 45.0 | 0.9712 | 273.9 | 176.8 | 0.596 | 4.005 | 0.436 | 682.3 | 617.9 |
| 50.0 | 0.9733 | 293.9 | 196.6 | 0.659 | 4.008 | 0.455 | 688.9 | 570.8 |
| 60.0 | 0.9780 | 334.0 | 236.2 | 0.781 | 4.012 | 0.494 | 700.8 | 492.8 |
| 70.0 | 0.9830 | 374.2 | 275.9 | 0.900 | 4.017 | 0.530 | 710.8 | 431.3 |
| 80.0 | 0.9884 | 414.4 | 315.5 | 1.015 | 4.023 | 0.565 | 719.1 | 381.9 |
| 90.0 | 0.9942 | 454.6 | 355.2 | 1.128 | 4.031 | 0.599 | 726.0 | 341.6 |
| 100.0 | 1.0003 | 495.0 | 395.0 | 1.237 | 4.039 | 0.632 | 731.6 | 308.4 |
| 125.0 | 1.0173 | 596.3 | 494.5 | 1.500 | 4.062 | 0.712 | 741.2 | 247.1 |
| 150.0 | 1.0365 | 698.1 | 594.4 | 1.748 | 4.084 | 0.790 | 745.6 | 206.0 |
| 175.0 | 1.0582 | 800.5 | 694.7 | 1.983 | 4.109 | 0.869 | 745.5 | 177.1 |
| 200.0 | 1.0826 | 903.6 | 795.3 | 2.207 | 4.141 | 0.952 | 740.7 | 156.0 |
| 225.0 | 1.1099 | 1007 | 896.6 | 2.421 | 4.182 | 1.043 | 731.3 | 140.0 |
| 250.0 | 1.1406 | 1112 | 998.7 | 2.627 | 4.236 | 1.143 | 717.2 | 127.5 |
| 275.0 | 1.1753 | 1219 | 1102 | 2.826 | 4.306 | 1.257 | 698.5 | 117.5 |
| 300.0 | 1.2148 | 1328 | 1206 | 3.020 | 4.391 | 1.386 | 675.3 | 109.1 |
| 325.0 | 1.2599 | 1439 | 1313 | 3.210 | 4.494 | 1.536 | 648.0 | 102.0 |
| 350.0 | 1.3120 | 1553 | 1421 | 3.396 | 4.615 | 1.709 | 616.8 | 95.73 |
| 375.0 | 1.3726 | 1670 | 1532 | 3.580 | 4.754 | 1.911 | 583.0 | 90.02 |
| 400.0 | 1.4439 | 1790 | 1646 | 3.763 | 4.911 | 2.144 | 547.9 | 84.71 |
| 450.0 | 1.6292 | 2045 | 1882 | 4.127 | 5.261 | 2.704 | 476.0 | 74.86 |
| 500.0 | 1.8932 | 2316 | 2126 | 4.490 | 5.557 | 3.284 | 394.8 | 66.07 |
| 550.0 | 2.2504 | 2595 | 2370 | 4.840 | 5.541 | 3.544 | 319.2 | 59.11 |
| 600.0 | 2.6743 | 2863 | 2595 | 5.156 | 5.122 | 3.288 | 271.8 | 54.67 |
| 650.0 | 3.1148 | 3105 | 2794 | 5.426 | 4.581 | 2.802 | 248.0 | 52.43 |
| 700.0 | 3.5416 | 3323 | 2968 | 5.655 | 4.129 | 2.349 | 236.2 | 51.61 |
| 750.0 | 3.9460 | 3520 | 3126 | 5.853 | 3.797 | 1.992 | 229.7 | 51.63 |
| 800.0 | 4.3285 | 3704 | 3271 | 6.029 | 3.561 | 1.721 | 225.5 | 52.12 |
| 850.0 | 4.6918 | 3877 | 3408 | 6.187 | 3.392 | 1.512 | 222.6 | 52.91 |
| 900.0 | 5.0390 | 4044 | 3540 | 6.332 | 3.269 | 1.349 | 220.2 | 53.86 |
| 950.0 | 5.3725 | 4205 | 3668 | 6.466 | 3.179 | 1.219 | 218.2 | 54.92 |
| 1000.0 | 5.6943 | 4362 | 3793 | 6.592 | 3.113 | 1.112 | 216.3 | 56.04 |

**Table 2. Thermodynamic and Transport Properties of Saturated Water and Steam**

**Tafel 2. Thermodynamische und Transportgrößen von gesättigtem Wasser und Dampf**

Part 1/Teil 1: $\rho, h, r, u, s, c_p, c_v$ (Temperature Table/Temperaturtafel)

Part 2/Teil 2: $\lambda, \eta, \nu, a, Pr, \sigma, b, \beta_v, \beta_s$ (Temperature Table/Temperaturtafel)

Part 3/Teil 3: $\alpha_p, \delta_h, \delta_T, \dfrac{dr}{dT}, \dfrac{dp}{dT}\bigg|_{sat}, \dfrac{d^2 p}{d T^2}\bigg|_{sat}, c_{sat}, c, \chi_T$ (Temperature Table/Temperaturtafel)

Part 4/Teil 4: $\rho, h, r, u, s, c_p, c_v$ (Pressure Table/Drucktafel)

Part 5/Teil 5: $\lambda, \eta, \nu, a, Pr, \sigma, b, \beta_v, \beta_s$ (Pressure Table/Drucktafel)

Part 6/Teil 6: $\alpha_p, \delta_h, \delta_T, \dfrac{dr}{dT}, \dfrac{dp}{dT}\bigg|_{sat}, \dfrac{d^2 p}{d T^2}\bigg|_{sat}, c_{sat}, c, \chi_T$ (Pressure Table/Drucktafel)

Table 2, part 1

| $t$ | $p$ | $\varrho'$ | $\varrho''$ | $h'$ | $h''$ | $r$ |
|---|---|---|---|---|---|---|
| °C | MPa | kg/m³ | | kJ/kg | | kJ/kg |
| 0.0[1] | 0.00061 | 999.8 | 0.0049 | −0.0416 | 2500.5 | 2500.6 |
| 1.0 | 0.00066 | 999.8 | 0.0052 | 4.1832 | 2502.4 | 2498.2 |
| 2.0 | 0.00071 | 999.9 | 0.0056 | 8.4010 | 2504.2 | 2495.8 |
| 3.0 | 0.00076 | 999.9 | 0.0060 | 12.613 | 2506.0 | 2493.4 |
| 4.0 | 0.00081 | 999.9 | 0.0064 | 16.819 | 2507.9 | 2491.1 |
| 5.0 | 0.00087 | 999.9 | 0.0068 | 21.021 | 2509.7 | 2488.7 |
| 6.0 | 0.00094 | 999.9 | 0.0073 | 25.220 | 2511.5 | 2486.3 |
| 7.0 | 0.00100 | 999.9 | 0.0078 | 29.415 | 2513.4 | 2484.0 |
| 8.0 | 0.00107 | 999.8 | 0.0083 | 33.608 | 2515.2 | 2481.6 |
| 9.0 | 0.00115 | 999.8 | 0.0088 | 37.799 | 2517.1 | 2479.3 |
| 10.0 | 0.00123 | 999.7 | 0.0094 | 41.988 | 2518.9 | 2476.9 |
| 11.0 | 0.00131 | 999.6 | 0.0100 | 46.175 | 2520.7 | 2474.5 |
| 12.0 | 0.00140 | 999.5 | 0.0107 | 50.362 | 2522.6 | 2472.2 |
| 13.0 | 0.00150 | 999.4 | 0.0114 | 54.547 | 2524.4 | 2469.8 |
| 14.0 | 0.00160 | 999.2 | 0.0121 | 58.732 | 2526.2 | 2467.5 |
| 15.0 | 0.00171 | 999.1 | 0.0128 | 62.917 | 2528.0 | 2465.1 |
| 16.0 | 0.00182 | 998.9 | 0.0136 | 67.101 | 2529.9 | 2462.8 |
| 17.0 | 0.00194 | 998.8 | 0.0145 | 71.285 | 2531.7 | 2460.4 |
| 18.0 | 0.00206 | 998.6 | 0.0154 | 75.468 | 2533.5 | 2458.1 |
| 19.0 | 0.00220 | 998.4 | 0.0163 | 79.652 | 2535.3 | 2455.7 |
| 20.0 | 0.00234 | 998.2 | 0.0173 | 83.835 | 2537.2 | 2453.3 |
| 21.0 | 0.00249 | 998.0 | 0.0183 | 88.019 | 2539.0 | 2451.0 |
| 22.0 | 0.00264 | 997.7 | 0.0194 | 92.202 | 2540.8 | 2448.6 |
| 23.0 | 0.00281 | 997.5 | 0.0206 | 96.386 | 2542.6 | 2446.2 |
| 24.0 | 0.00299 | 997.3 | 0.0218 | 100.57 | 2544.5 | 2443.9 |
| 25.0 | 0.00317 | 997.0 | 0.0231 | 104.75 | 2546.3 | 2441.5 |
| 26.0 | 0.00336 | 996.8 | 0.0244 | 108.94 | 2548.1 | 2439.2 |
| 27.0 | 0.00357 | 996.5 | 0.0258 | 113.12 | 2549.9 | 2436.8 |
| 28.0 | 0.00378 | 996.2 | 0.0273 | 117.30 | 2551.7 | 2434.4 |
| 29.0 | 0.00401 | 995.9 | 0.0288 | 121.49 | 2553.5 | 2432.0 |

[1] The values of this line relate to metastable states.

| $u'$ | $u''$ | $s'$ | $s''$ | $c_p'$ | $c_p''$ | $c_v'$ | $c_v''$ |
|---|---|---|---|---|---|---|---|
| kJ/kg | | kJ/kg K | | kJ/kg K | | kJ/kg K | |
| −0.0423 | 2374.5 | 0.000 | 9.154 | 4.229 | 1.868 | 4.225 | 1.404 |
| 4.1826 | 2375.9 | 0.015 | 9.128 | 4.221 | 1.868 | 4.219 | 1.404 |
| 8.4003 | 2377.3 | 0.031 | 9.101 | 4.215 | 1.869 | 4.214 | 1.405 |
| 12.612 | 2378.7 | 0.046 | 9.075 | 4.209 | 1.869 | 4.209 | 1.405 |
| 16.818 | 2380.0 | 0.061 | 9.049 | 4.204 | 1.870 | 4.204 | 1.406 |
| 21.020 | 2381.4 | 0.076 | 9.024 | 4.200 | 1.871 | 4.200 | 1.406 |
| 25.219 | 2382.8 | 0.091 | 8.998 | 4.197 | 1.871 | 4.196 | 1.407 |
| 29.414 | 2384.2 | 0.106 | 8.973 | 4.194 | 1.872 | 4.193 | 1.407 |
| 33.607 | 2385.6 | 0.121 | 8.948 | 4.192 | 1.872 | 4.190 | 1.408 |
| 37.798 | 2386.9 | 0.136 | 8.923 | 4.190 | 1.873 | 4.187 | 1.408 |
| 41.986 | 2388.3 | 0.151 | 8.899 | 4.188 | 1.874 | 4.184 | 1.409 |
| 46.174 | 2389.7 | 0.166 | 8.874 | 4.187 | 1.875 | 4.181 | 1.409 |
| 50.360 | 2391.1 | 0.180 | 8.850 | 4.186 | 1.875 | 4.178 | 1.410 |
| 54.546 | 2392.4 | 0.195 | 8.826 | 4.185 | 1.876 | 4.175 | 1.411 |
| 58.731 | 2393.8 | 0.210 | 8.803 | 4.185 | 1.877 | 4.173 | 1.411 |
| 62.915 | 2395.2 | 0.224 | 8.779 | 4.184 | 1.878 | 4.170 | 1.412 |
| 67.099 | 2396.6 | 0.239 | 8.756 | 4.184 | 1.879 | 4.167 | 1.412 |
| 71.283 | 2397.9 | 0.253 | 8.733 | 4.184 | 1.879 | 4.164 | 1.413 |
| 75.466 | 2399.3 | 0.268 | 8.710 | 4.183 | 1.880 | 4.161 | 1.414 |
| 79.650 | 2400.7 | 0.282 | 8.688 | 4.183 | 1.881 | 4.158 | 1.414 |
| 83.833 | 2402.0 | 0.296 | 8.665 | 4.183 | 1.882 | 4.155 | 1.415 |
| 88.016 | 2403.4 | 0.310 | 8.643 | 4.183 | 1.883 | 4.152 | 1.416 |
| 92.200 | 2404.8 | 0.325 | 8.621 | 4.183 | 1.884 | 4.149 | 1.416 |
| 96.383 | 2406.1 | 0.339 | 8.599 | 4.183 | 1.885 | 4.146 | 1.417 |
| 100.57 | 2407.5 | 0.353 | 8.577 | 4.183 | 1.886 | 4.142 | 1.418 |
| 104.75 | 2408.9 | 0.367 | 8.556 | 4.183 | 1.887 | 4.139 | 1.419 |
| 108.93 | 2410.2 | 0.381 | 8.535 | 4.183 | 1.888 | 4.135 | 1.420 |
| 113.12 | 2411.6 | 0.395 | 8.513 | 4.183 | 1.889 | 4.132 | 1.420 |
| 117.30 | 2413.0 | 0.409 | 8.493 | 4.183 | 1.890 | 4.128 | 1.421 |
| 121.48 | 2414.3 | 0.423 | 8.472 | 4.183 | 1.891 | 4.124 | 1.422 |

(1) Die Werte dieser Zeile gelten für metastabile Zustände.

Table 2, part 1 (cont.)

| t | p | $\varrho'$ | $\varrho''$ | $h'$ | $h''$ | r |
|---|---|---|---|---|---|---|
| °C | MPa | kg/m³ | | kJ/kg | | kJ/kg |
| 30.0 | 0.00425 | 995.6 | 0.0304 | 125.67 | 2555.3 | 2429.7 |
| 32.5 | 0.00489 | 994.8 | 0.0348 | 136.13 | 2559.9 | 2423.7 |
| 35.0 | 0.00563 | 994.0 | 0.0397 | 146.59 | 2564.4 | 2417.8 |
| 37.5 | 0.00645 | 993.1 | 0.0451 | 157.05 | 2568.9 | 2411.8 |
| 40.0 | 0.00738 | 992.2 | 0.0512 | 167.50 | 2573.4 | 2405.9 |
| 42.5 | 0.00842 | 991.2 | 0.0580 | 177.96 | 2577.8 | 2399.9 |
| 45.0 | 0.00959 | 990.2 | 0.0655 | 188.42 | 2582.3 | 2393.9 |
| 47.5 | 0.01089 | 989.1 | 0.0739 | 198.87 | 2586.8 | 2387.9 |
| 50.0 | 0.01234 | 988.0 | 0.0831 | 209.33 | 2591.2 | 2381.9 |
| 52.5 | 0.01396 | 986.8 | 0.0933 | 219.78 | 2595.6 | 2375.8 |
| 55.0 | 0.01575 | 985.7 | 0.1045 | 230.24 | 2600.0 | 2369.8 |
| 57.5 | 0.01774 | 984.4 | 0.1168 | 240.70 | 2604.4 | 2363.7 |
| 60.0 | 0.01993 | 983.2 | 0.1303 | 251.15 | 2608.8 | 2357.6 |
| 62.5 | 0.02235 | 981.9 | 0.1451 | 261.61 | 2613.1 | 2351.5 |
| 65.0 | 0.02502 | 980.5 | 0.1613 | 272.08 | 2617.5 | 2345.4 |
| 67.5 | 0.02796 | 979.2 | 0.1790 | 282.54 | 2621.8 | 2339.3 |
| 70.0 | 0.03118 | 977.7 | 0.1982 | 293.01 | 2626.1 | 2333.1 |
| 72.5 | 0.03470 | 976.3 | 0.2192 | 303.48 | 2630.4 | 2326.9 |
| 75.0 | 0.03856 | 974.8 | 0.2419 | 313.96 | 2634.6 | 2320.7 |
| 77.5 | 0.04278 | 973.3 | 0.2666 | 324.44 | 2638.9 | 2314.4 |
| 80.0 | 0.04737 | 971.8 | 0.2934 | 334.93 | 2643.1 | 2308.1 |
| 82.5 | 0.05238 | 970.2 | 0.3223 | 345.42 | 2647.3 | 2301.8 |
| 85.0 | 0.05781 | 968.6 | 0.3535 | 355.92 | 2651.4 | 2295.5 |
| 87.5 | 0.06372 | 967.0 | 0.3872 | 366.42 | 2655.5 | 2289.1 |
| 90.0 | 0.07012 | 965.3 | 0.4234 | 376.93 | 2659.6 | 2282.7 |
| 92.5 | 0.07704 | 963.6 | 0.4624 | 387.45 | 2663.7 | 2276.2 |
| 95.0 | 0.08453 | 961.9 | 0.5043 | 397.98 | 2667.7 | 2269.8 |
| 97.5 | 0.09261 | 960.2 | 0.5493 | 408.52 | 2671.8 | 2263.2 |
| 99.63 | 0.10000 | 958.7 | 0.5902 | 417.51 | 2675.1 | 2257.6 |
| 100.0 | 0.10130 | 958.4 | 0.5975 | 419.06 | 2675.7 | 2256.7 |

| $u'$ | $u''$ | $s'$ | $s''$ | $c_p'$ | $c_p''$ | $c_v'$ | $c_v''$ |
|---|---|---|---|---|---|---|---|
| kJ/kg | | kJ/kg K | | kJ/kg K | | kJ/kg K | |
| 125.67 | 2415.7 | 0.437 | 8.451 | 4.183 | 1.892 | 4.120 | 1.423 |
| 136.12 | 2419.1 | 0.471 | 8.401 | 4.183 | 1.895 | 4.110 | 1.425 |
| 146.58 | 2422.5 | 0.505 | 8.351 | 4.183 | 1.898 | 4.099 | 1.427 |
| 157.04 | 2425.8 | 0.539 | 8.303 | 4.183 | 1.901 | 4.088 | 1.430 |
| 167.50 | 2429.2 | 0.572 | 8.255 | 4.182 | 1.904 | 4.076 | 1.432 |
| 177.95 | 2432.6 | 0.606 | 8.209 | 4.182 | 1.908 | 4.064 | 1.435 |
| 188.41 | 2435.9 | 0.639 | 8.163 | 4.182 | 1.912 | 4.052 | 1.437 |
| 198.86 | 2439.3 | 0.671 | 8.118 | 4.182 | 1.915 | 4.040 | 1.440 |
| 209.31 | 2442.6 | 0.704 | 8.075 | 4.182 | 1.919 | 4.027 | 1.443 |
| 219.77 | 2445.9 | 0.736 | 8.032 | 4.182 | 1.924 | 4.014 | 1.446 |
| 230.22 | 2449.2 | 0.768 | 7.990 | 4.182 | 1.928 | 4.002 | 1.449 |
| 240.68 | 2452.5 | 0.800 | 7.948 | 4.182 | 1.932 | 3.989 | 1.452 |
| 251.13 | 2455.8 | 0.831 | 7.908 | 4.183 | 1.937 | 3.976 | 1.455 |
| 261.59 | 2459.1 | 0.862 | 7.868 | 4.184 | 1.942 | 3.963 | 1.459 |
| 272.05 | 2462.4 | 0.894 | 7.830 | 4.184 | 1.947 | 3.950 | 1.463 |
| 282.51 | 2465.6 | 0.924 | 7.791 | 4.186 | 1.953 | 3.937 | 1.466 |
| 292.98 | 2468.8 | 0.955 | 7.754 | 4.187 | 1.958 | 3.924 | 1.470 |
| 303.45 | 2472.0 | 0.985 | 7.717 | 4.188 | 1.964 | 3.911 | 1.474 |
| 313.92 | 2475.2 | 1.016 | 7.681 | 4.190 | 1.970 | 3.898 | 1.478 |
| 324.40 | 2478.4 | 1.046 | 7.646 | 4.192 | 1.976 | 3.885 | 1.482 |
| 334.88 | 2481.6 | 1.075 | 7.611 | 4.194 | 1.983 | 3.872 | 1.486 |
| 345.36 | 2484.7 | 1.105 | 7.577 | 4.197 | 1.989 | 3.859 | 1.491 |
| 355.86 | 2487.9 | 1.134 | 7.544 | 4.199 | 1.996 | 3.846 | 1.496 |
| 366.36 | 2491.0 | 1.164 | 7.511 | 4.202 | 2.004 | 3.834 | 1.500 |
| 376.86 | 2494.0 | 1.193 | 7.478 | 4.204 | 2.011 | 3.821 | 1.505 |
| 387.37 | 2497.1 | 1.221 | 7.447 | 4.207 | 2.019 | 3.808 | 1.510 |
| 397.89 | 2500.1 | 1.250 | 7.415 | 4.210 | 2.027 | 3.796 | 1.515 |
| 408.42 | 2503.2 | 1.279 | 7.385 | 4.214 | 2.035 | 3.783 | 1.521 |
| 417.41 | 2505.7 | 1.303 | 7.359 | 4.217 | 2.043 | 3.773 | 1.525 |
| 418.96 | 2506.1 | 1.307 | 7.354 | 4.217 | 2.044 | 3.771 | 1.526 |

Table 2, part 1 (cont.)

| t | p | ϱ' | ϱ'' | h' | h'' | r |
|---|---|---|---|---|---|---|
| °C | MPa | kg/m³ | | kJ/kg | | kJ/kg |
| 105.0 | 0.1208 | 954.7 | 0.7042 | 440.18 | 2683.6 | 2243.4 |
| 110.0 | 0.1432 | 951.0 | 0.8260 | 461.34 | 2691.3 | 2229.9 |
| 115.0 | 0.1690 | 947.1 | 0.9643 | 482.54 | 2698.8 | 2216.3 |
| 120.0 | 0.1985 | 943.2 | 1.1208 | 503.78 | 2706.2 | 2202.4 |
| 125.0 | 0.2320 | 939.1 | 1.2972 | 525.07 | 2713.4 | 2188.3 |
| 130.0 | 0.2700 | 934.9 | 1.4954 | 546.41 | 2720.4 | 2174.0 |
| 135.0 | 0.3129 | 930.6 | 1.7172 | 567.80 | 2727.2 | 2159.4 |
| 140.0 | 0.3612 | 926.2 | 1.9647 | 589.24 | 2733.8 | 2144.6 |
| 145.0 | 0.4153 | 921.7 | 2.2400 | 610.75 | 2740.2 | 2129.5 |
| 150.0 | 0.4757 | 917.1 | 2.5454 | 632.32 | 2746.4 | 2114.1 |
| 155.0 | 0.5430 | 912.3 | 2.8834 | 653.95 | 2752.3 | 2098.3 |
| 160.0 | 0.6177 | 907.5 | 3.2564 | 675.65 | 2758.0 | 2082.3 |
| 165.0 | 0.7003 | 902.6 | 3.6670 | 697.43 | 2763.3 | 2065.9 |
| 170.0 | 0.7915 | 897.5 | 4.1181 | 719.28 | 2768.5 | 2049.2 |
| 175.0 | 0.8918 | 892.3 | 4.6127 | 741.22 | 2773.3 | 2032.0 |
| 180.0 | 1.0019 | 887.1 | 5.1539 | 763.25 | 2777.8 | 2014.5 |
| 185.0 | 1.1225 | 881.7 | 5.7450 | 785.37 | 2782.0 | 1996.6 |
| 190.0 | 1.2542 | 876.1 | 6.3896 | 807.60 | 2785.8 | 1978.2 |
| 195.0 | 1.3976 | 870.5 | 7.0913 | 829.93 | 2789.4 | 1959.4 |
| 200.0 | 1.5536 | 864.7 | 7.8542 | 852.38 | 2792.5 | 1940.1 |
| 205.0 | 1.7229 | 858.8 | 8.6825 | 874.96 | 2795.3 | 1920.4 |
| 210.0 | 1.9062 | 852.8 | 9.5807 | 897.66 | 2797.7 | 1900.0 |
| 215.0 | 2.1042 | 846.6 | 10.554 | 920.51 | 2799.7 | 1879.2 |
| 220.0 | 2.3178 | 840.3 | 11.607 | 943.51 | 2801.3 | 1857.8 |
| 225.0 | 2.5479 | 833.9 | 12.745 | 966.67 | 2802.4 | 1835.7 |
| 230.0 | 2.7951 | 827.3 | 13.976 | 990.00 | 2803.1 | 1813.1 |
| 235.0 | 3.0604 | 820.5 | 15.304 | 1013.5 | 2803.3 | 1789.7 |
| 240.0 | 3.3447 | 813.5 | 16.739 | 1037.2 | 2803.0 | 1765.7 |
| 245.0 | 3.6488 | 806.4 | 18.286 | 1061.2 | 2802.1 | 1741.0 |
| 250.0 | 3.9736 | 799.1 | 19.956 | 1085.3 | 2800.7 | 1715.4 |

| $u'$ | $u''$ | $s'$ | $s''$ | $c_p'$ | $c_p''$ | $c_v'$ | $c_v''$ |
|---|---|---|---|---|---|---|---|
| kJ/kg | | kJ/kg K | | kJ/kg K | | kJ/kg K | |
| 440.05 | 2512.1 | 1.363 | 7.296 | 4.224 | 2.062 | 3.746 | 1.538 |
| 461.19 | 2517.9 | 1.419 | 7.239 | 4.232 | 2.082 | 3.721 | 1.550 |
| 482.36 | 2523.5 | 1.473 | 7.183 | 4.240 | 2.103 | 3.697 | 1.562 |
| 503.57 | 2529.1 | 1.528 | 7.130 | 4.249 | 2.126 | 3.672 | 1.576 |
| 524.82 | 2534.5 | 1.581 | 7.078 | 4.258 | 2.150 | 3.648 | 1.590 |
| 546.12 | 2539.8 | 1.635 | 7.027 | 4.267 | 2.176 | 3.624 | 1.605 |
| 567.46 | 2545.0 | 1.687 | 6.978 | 4.278 | 2.203 | 3.600 | 1.621 |
| 588.85 | 2550.0 | 1.739 | 6.930 | 4.288 | 2.233 | 3.576 | 1.638 |
| 610.30 | 2554.8 | 1.791 | 6.884 | 4.300 | 2.265 | 3.553 | 1.656 |
| 631.80 | 2559.5 | 1.842 | 6.838 | 4.312 | 2.299 | 3.530 | 1.674 |
| 653.35 | 2564.0 | 1.893 | 6.794 | 4.325 | 2.335 | 3.507 | 1.694 |
| 674.97 | 2568.3 | 1.943 | 6.750 | 4.339 | 2.374 | 3.484 | 1.715 |
| 696.65 | 2572.4 | 1.993 | 6.708 | 4.353 | 2.415 | 3.462 | 1.736 |
| 718.40 | 2576.3 | 2.042 | 6.666 | 4.369 | 2.460 | 3.440 | 1.759 |
| 740.22 | 2579.9 | 2.091 | 6.625 | 4.385 | 2.507 | 3.419 | 1.783 |
| 762.12 | 2583.4 | 2.140 | 6.585 | 4.403 | 2.558 | 3.397 | 1.809 |
| 784.10 | 2586.6 | 2.188 | 6.546 | 4.423 | 2.612 | 3.377 | 1.835 |
| 806.17 | 2589.6 | 2.236 | 6.507 | 4.443 | 2.670 | 3.356 | 1.863 |
| 828.33 | 2592.3 | 2.283 | 6.469 | 4.465 | 2.731 | 3.337 | 1.892 |
| 850.58 | 2594.7 | 2.331 | 6.431 | 4.489 | 2.797 | 3.317 | 1.923 |
| 872.95 | 2596.9 | 2.378 | 6.394 | 4.515 | 2.867 | 3.298 | 1.954 |
| 895.43 | 2598.7 | 2.425 | 6.357 | 4.542 | 2.943 | 3.280 | 1.988 |
| 918.02 | 2600.3 | 2.471 | 6.321 | 4.572 | 3.023 | 3.262 | 2.022 |
| 940.75 | 2601.6 | 2.518 | 6.285 | 4.604 | 3.109 | 3.245 | 2.058 |
| 963.61 | 2602.5 | 2.564 | 6.249 | 4.638 | 3.201 | 3.228 | 2.096 |
| 986.62 | 2603.1 | 2.610 | 6.213 | 4.675 | 3.299 | 3.212 | 2.135 |
| 1009.8 | 2603.3 | 2.656 | 6.178 | 4.715 | 3.405 | 3.197 | 2.175 |
| 1033.1 | 2603.1 | 2.701 | 6.142 | 4.759 | 3.519 | 3.182 | 2.217 |
| 1056.6 | 2602.6 | 2.747 | 6.107 | 4.806 | 3.641 | 3.168 | 2.261 |
| 1080.3 | 2601.6 | 2.793 | 6.072 | 4.857 | 3.772 | 3.154 | 2.306 |

| $t$ | $p$ | $\varrho'$ | $\varrho''$ | $h'$ | $h''$ | $r$ |
|---|---|---|---|---|---|---|
| °C | MPa | kg/m³ | | kJ/kg | | kJ/kg |
| 255.0 | 4.3202 | 791.6 | 21.757 | 1109.7 | 2798.8 | 1689.1 |
| 260.0 | 4.6894 | 783.8 | 23.700 | 1134.4 | 2796.2 | 1661.9 |
| 265.0 | 5.0823 | 775.9 | 25.797 | 1159.3 | 2793.0 | 1633.7 |
| 270.0 | 5.4999 | 767.7 | 28.061 | 1184.6 | 2789.1 | 1604.6 |
| 275.0 | 5.9431 | 759.2 | 30.507 | 1210.1 | 2784.5 | 1574.4 |
| 280.0 | 6.4132 | 750.5 | 33.152 | 1236.1 | 2779.2 | 1543.1 |
| 285.0 | 6.9111 | 741.5 | 36.015 | 1262.4 | 2773.0 | 1510.6 |
| 290.0 | 7.4380 | 732.2 | 39.119 | 1289.1 | 2765.9 | 1476.7 |
| 295.0 | 7.9952 | 722.5 | 42.488 | 1316.3 | 2757.8 | 1441.5 |
| 300.0 | 8.5838 | 712.4 | 46.154 | 1344.1 | 2748.7 | 1404.7 |
| 305.0 | 9.2051 | 701.9 | 50.152 | 1372.3 | 2738.5 | 1366.2 |
| 310.0 | 9.8605 | 691.0 | 54.525 | 1401.2 | 2727.0 | 1325.8 |
| 315.0 | 10.551 | 679.5 | 59.324 | 1430.8 | 2714.2 | 1283.3 |
| 320.0 | 11.279 | 667.4 | 64.615 | 1461.3 | 2699.7 | 1238.5 |
| 325.0 | 12.046 | 654.6 | 70.477 | 1492.6 | 2683.5 | 1190.9 |
| 330.0 | 12.852 | 641.0 | 77.013 | 1525.0 | 2665.3 | 1140.3 |
| 335.0 | 13.701 | 626.5 | 84.357 | 1558.6 | 2644.7 | 1086.1 |
| 340.0 | 14.594 | 610.8 | 92.691 | 1593.8 | 2621.3 | 1027.5 |
| 345.0 | 15.533 | 593.7 | 102.27 | 1630.9 | 2594.5 | 963.64 |
| 350.0 | 16.521 | 574.7 | 113.48 | 1670.4 | 2563.5 | 893.03 |
| 355.0 | 17.561 | 553.2 | 126.92 | 1713.3 | 2526.7 | 813.46 |
| 360.0 | 18.655 | 528.1 | 143.64 | 1761.0 | 2482.0 | 721.06 |
| 365.0 | 19.809 | 497.0 | 165.88 | 1816.7 | 2424.6 | 607.90 |
| 370.0 | 21.030 | 453.1 | 200.29 | 1889.7 | 2340.2 | 450.42 |
| 371.0 | 21.283 | 440.7 | 210.63 | 1909.3 | 2315.8 | 406.51 |
| 372.0 | 21.539 | 425.8 | 224.55 | 1932.4 | 2283.4 | 351.04 |
| 373.0 | 21.799 | 403.8 | 244.67 | 1964.8 | 2238.1 | 273.37 |
| 374.0 | 22.055 | 322.0 | | 2085.8 | | 0.00 |

| $u'$ | $u''$ | $s'$ | $s''$ | $c_p'$ | $c_p''$ | $c_v'$ | $c_v''$ |
|------|-------|------|-------|--------|---------|--------|---------|
| kJ/kg | | kJ/kg K | | kJ/kg K | | kJ/kg K | |
| 1104.3 | 2600.2 | 2.838 | 6.036 | 4.912 | 3.915 | 3.141 | 2.353 |
| 1128.4 | 2598.4 | 2.884 | 6.001 | 4.973 | 4.068 | 3.129 | 2.401 |
| 1152.8 | 2596.0 | 2.929 | 5.965 | 5.039 | 4.236 | 3.117 | 2.451 |
| 1177.4 | 2593.2 | 2.975 | 5.929 | 5.111 | 4.418 | 3.107 | 2.503 |
| 1202.3 | 2589.7 | 3.021 | 5.893 | 5.191 | 4.617 | 3.097 | 2.557 |
| 1227.5 | 2585.7 | 3.067 | 5.857 | 5.279 | 4.836 | 3.088 | 2.612 |
| 1253.1 | 2581.1 | 3.113 | 5.819 | 5.377 | 5.077 | 3.080 | 2.669 |
| 1279.0 | 2575.7 | 3.159 | 5.782 | 5.485 | 5.345 | 3.073 | 2.728 |
| 1305.3 | 2569.7 | 3.206 | 5.743 | 5.608 | 5.644 | 3.067 | 2.788 |
| 1332.0 | 2562.8 | 3.253 | 5.704 | 5.746 | 5.981 | 3.062 | 2.851 |
| 1359.2 | 2555.0 | 3.301 | 5.664 | 5.903 | 6.362 | 3.058 | 2.916 |
| 1387.0 | 2546.2 | 3.349 | 5.623 | 6.084 | 6.799 | 3.056 | 2.983 |
| 1415.3 | 2536.3 | 3.398 | 5.580 | 6.294 | 7.305 | 3.056 | 3.053 |
| 1444.4 | 2525.2 | 3.448 | 5.536 | 6.542 | 7.898 | 3.057 | 3.125 |
| 1474.2 | 2512.6 | 3.498 | 5.489 | 6.839 | 8.603 | 3.061 | 3.200 |
| 1504.9 | 2498.4 | 3.550 | 5.441 | 7.201 | 9.458 | 3.068 | 3.277 |
| 1536.8 | 2482.3 | 3.603 | 5.389 | 7.654 | 10.52 | 3.077 | 3.358 |
| 1569.9 | 2463.9 | 3.659 | 5.335 | 8.238 | 11.87 | 3.091 | 3.442 |
| 1604.7 | 2442.7 | 3.716 | 5.275 | 9.020 | 13.65 | 3.110 | 3.530 |
| 1641.7 | 2417.9 | 3.777 | 5.210 | 10.13 | 16.11 | 3.136 | 3.622 |
| 1681.5 | 2388.4 | 3.843 | 5.138 | 11.81 | 19.77 | 3.172 | 3.718 |
| 1725.6 | 2352.2 | 3.915 | 5.054 | 14.69 | 25.80 | 3.222 | 3.819 |
| 1776.8 | 2305.2 | 3.999 | 4.952 | 20.73 | 37.80 | 3.298 | 3.915 |
| 1843.3 | 2235.2 | 4.109 | 4.810 | 41.96 | 78.75 | 3.443 | 4.065 |
| 1861.0 | 2214.8 | 4.139 | 4.770 | 55.58 | 104 | 3.503 | 4.128 |
| 1881.8 | 2187.5 | 4.174 | 4.718 | 84.06 | 167 | 3.619 | 4.280 |
| 1910.8 | 2149.0 | 4.223 | 4.646 | 197 | 411 | 4.008 | 4.747 |
| 2017.3 | | 4.409 | | $\infty$ | | — | |

Table 2, part 2

| t | p | λ′ | λ″ | η′ | η″ | ν′ | ν″ |
|---|---|---|---|---|---|---|---|
| °C | MPa | mW/Km | | μPa s | | $10^{-6}$ m²/s | |
| 0.0[1] | 0.00061 | 561.0 | 17.07 | 1793 | 9.216 | 1.793 | 1900 |
| 1.0 | 0.00066 | 562.9 | 17.12 | 1732 | 9.239 | 1.732 | 1778 |
| 2.0 | 0.00071 | 564.8 | 17.18 | 1674 | 9.263 | 1.674 | 1665 |
| 3.0 | 0.00076 | 566.7 | 17.23 | 1620 | 9.287 | 1.620 | 1560 |
| 4.0 | 0.00081 | 568.6 | 17.28 | 1568 | 9.311 | 1.568 | 1463 |
| 5.0 | 0.00087 | 570.5 | 17.34 | 1519 | 9.335 | 1.519 | 1373 |
| 6.0 | 0.00094 | 572.4 | 17.39 | 1472 | 9.360 | 1.472 | 1288 |
| 7.0 | 0.00100 | 574.3 | 17.45 | 1428 | 9.385 | 1.428 | 1210 |
| 8.0 | 0.00107 | 576.2 | 17.51 | 1385 | 9.410 | 1.386 | 1137 |
| 9.0 | 0.00115 | 578.1 | 17.56 | 1345 | 9.435 | 1.345 | 1069 |
| 10.0 | 0.00123 | 580.0 | 17.62 | 1306 | 9.461 | 1.307 | 1006 |
| 11.0 | 0.00131 | 581.9 | 17.68 | 1270 | 9.486 | 1.270 | 946.8 |
| 12.0 | 0.00140 | 583.8 | 17.74 | 1235 | 9.512 | 1.235 | 891.7 |
| 13.0 | 0.00150 | 585.6 | 17.80 | 1201 | 9.538 | 1.202 | 840.2 |
| 14.0 | 0.00160 | 587.5 | 17.86 | 1169 | 9.565 | 1.170 | 792.1 |
| 15.0 | 0.00171 | 589.3 | 17.92 | 1138 | 9.591 | 1.139 | 747.1 |
| 16.0 | 0.00182 | 591.2 | 17.98 | 1109 | 9.618 | 1.110 | 705.1 |
| 17.0 | 0.00194 | 593.0 | 18.04 | 1080 | 9.645 | 1.082 | 665.7 |
| 18.0 | 0.00206 | 594.8 | 18.10 | 1053 | 9.672 | 1.055 | 628.9 |
| 19.0 | 0.00220 | 596.6 | 18.16 | 1027 | 9.699 | 1.029 | 594.3 |
| 20.0 | 0.00234 | 598.4 | 18.23 | 1002 | 9.727 | 1.004 | 562.0 |
| 21.0 | 0.00249 | 600.2 | 18.29 | 978.0 | 9.754 | 0.9800 | 531.6 |
| 22.0 | 0.00264 | 601.9 | 18.35 | 954.8 | 9.782 | 0.9570 | 503.2 |
| 23.0 | 0.00281 | 603.7 | 18.42 | 932.6 | 9.810 | 0.9349 | 476.4 |
| 24.0 | 0.00299 | 605.4 | 18.48 | 911.1 | 9.838 | 0.9136 | 451.4 |
| 25.0 | 0.00317 | 607.1 | 18.55 | 890.5 | 9.866 | 0.8931 | 427.8 |
| 26.0 | 0.00336 | 608.8 | 18.62 | 870.6 | 9.895 | 0.8734 | 405.6 |
| 27.0 | 0.00357 | 610.5 | 18.68 | 851.4 | 9.923 | 0.8544 | 384.8 |
| 28.0 | 0.00378 | 612.2 | 18.75 | 832.8 | 9.952 | 0.8360 | 365.1 |
| 29.0 | 0.00401 | 613.8 | 18.82 | 815.0 | 9.981 | 0.8183 | 346.7 |

(1) The values of this line relate to metastable states.

| $a'$ | $a''$ | $Pr'$ | $Pr''$ | $\sigma$ | $b$ | $\beta_v'$ | $\beta_s'$ |
|------|------|------|------|------|------|------|------|
| $10^{-6}$ m²/s | | — | | mN/m | mm | 1/K | |
| 0.1327 | 1884 | 13.51 | 1.008 | 75.65 | 2.778 | −258.6 | −314000 |
| 0.1334 | 1764 | 12.99 | 1.008 | 75.51 | 2.775 | −182.8 | −385410 |
| 0.1340 | 1652 | 12.49 | 1.008 | 75.37 | 2.772 | −117.7 | −519670 |
| 0.1347 | 1549 | 12.03 | 1.008 | 75.23 | 2.770 | −62.06 | −856950 |
| 0.1353 | 1452 | 11.59 | 1.007 | 75.09 | 2.767 | −14.60 | −3170970 |
| 0.1358 | 1363 | 11.18 | 1.007 | 74.95 | 2.765 | 25.73 | 1567100 |
| 0.1364 | 1279 | 10.79 | 1.007 | 74.80 | 2.762 | 59.83 | 587640 |
| 0.1370 | 1202 | 10.43 | 1.007 | 74.66 | 2.759 | 88.51 | 346710 |
| 0.1375 | 1130 | 10.08 | 1.007 | 74.51 | 2.757 | 112.5 | 238370 |
| 0.1380 | 1062 | 9.747 | 1.006 | 74.37 | 2.754 | 132.3 | 177170 |
| 0.1385 | 999.8 | 9.434 | 1.006 | 74.22 | 2.752 | 148.6 | 138080 |
| 0.1390 | 941.2 | 9.136 | 1.006 | 74.08 | 2.749 | 161.8 | 111100 |
| 0.1395 | 886.6 | 8.853 | 1.006 | 73.93 | 2.746 | 172.3 | 91481 |
| 0.1400 | 835.6 | 8.583 | 1.006 | 73.79 | 2.744 | 180.5 | 76646 |
| 0.1405 | 787.9 | 8.325 | 1.005 | 73.64 | 2.741 | 186.7 | 65097 |
| 0.1410 | 743.3 | 8.080 | 1.005 | 73.49 | 2.739 | 191.1 | 55899 |
| 0.1414 | 701.6 | 7.845 | 1.005 | 73.34 | 2.736 | 194.0 | 48436 |
| 0.1419 | 662.5 | 7.621 | 1.005 | 73.19 | 2.734 | 195.7 | 42290 |
| 0.1424 | 625.9 | 7.407 | 1.005 | 73.04 | 2.731 | 196.3 | 37162 |
| 0.1428 | 591.7 | 7.202 | 1.005 | 72.89 | 2.729 | 195.9 | 32838 |
| 0.1433 | 559.6 | 7.005 | 1.004 | 72.74 | 2.726 | 194.7 | 29159 |
| 0.1438 | 529.4 | 6.817 | 1.004 | 72.59 | 2.723 | 192.9 | 26004 |
| 0.1442 | 501.2 | 6.636 | 1.004 | 72.44 | 2.721 | 190.5 | 23278 |
| 0.1447 | 474.6 | 6.462 | 1.004 | 72.28 | 2.718 | 187.7 | 20909 |
| 0.1451 | 449.7 | 6.296 | 1.004 | 72.13 | 2.716 | 184.5 | 18840 |
| 0.1456 | 426.2 | 6.136 | 1.004 | 71.98 | 2.713 | 180.9 | 17023 |
| 0.1460 | 404.2 | 5.982 | 1.003 | 71.82 | 2.711 | 177.1 | 15420 |
| 0.1464 | 383.5 | 5.834 | 1.003 | 71.67 | 2.708 | 173.1 | 14001 |
| 0.1469 | 364.0 | 5.691 | 1.003 | 71.51 | 2.706 | 169.0 | 12740 |
| 0.1473 | 345.6 | 5.554 | 1.003 | 71.35 | 2.703 | 164.8 | 11616 |

(1) Die Werte dieser Zeile gelten für metastabile Zustände.

Table 2, part 2 (cont.)

| $t$ | $p$ | $\lambda'$ | $\lambda''$ | $\eta'$ | $\eta''$ | $\nu'$ | $\nu''$ |
|------|---------|------|-------|-------|-------|--------|--------|
| °C | MPa | mW/Km | | $\mu$Pa s | | $10^{-6}$ m²/s | |
| 30.0 | 0.00425 | 615.4 | 18.89 | 797.7 | 10.01 | 0.8012 | 329.3 |
| 32.5 | 0.00489 | 619.4 | 19.06 | 757.0 | 10.08 | 0.7610 | 290.1 |
| 35.0 | 0.00563 | 623.2 | 19.23 | 719.6 | 10.16 | 0.7240 | 256.2 |
| 37.5 | 0.00645 | 627.0 | 19.41 | 685.1 | 10.23 | 0.6899 | 226.8 |
| 40.0 | 0.00738 | 630.5 | 19.60 | 653.2 | 10.31 | 0.6584 | 201.3 |
| 42.5 | 0.00842 | 634.0 | 19.78 | 623.7 | 10.38 | 0.6293 | 179.1 |
| 45.0 | 0.00959 | 637.3 | 19.97 | 596.3 | 10.46 | 0.6022 | 159.7 |
| 47.5 | 0.01089 | 640.5 | 20.17 | 570.8 | 10.54 | 0.5771 | 142.7 |
| 50.0 | 0.01234 | 643.5 | 20.36 | 547.1 | 10.62 | 0.5537 | 127.8 |
| 52.5 | 0.01396 | 646.4 | 20.56 | 524.9 | 10.69 | 0.5319 | 114.7 |
| 55.0 | 0.01575 | 649.2 | 20.77 | 504.2 | 10.77 | 0.5115 | 103.1 |
| 57.5 | 0.01774 | 651.8 | 20.97 | 484.8 | 10.85 | 0.4924 | 92.93 |
| 60.0 | 0.01993 | 654.3 | 21.18 | 466.6 | 10.93 | 0.4746 | 83.91 |
| 62.5 | 0.02235 | 656.7 | 21.40 | 449.5 | 11.01 | 0.4578 | 75.90 |
| 65.0 | 0.02502 | 658.9 | 21.62 | 433.4 | 11.10 | 0.4420 | 68.79 |
| 67.5 | 0.02796 | 661.1 | 21.84 | 418.3 | 11.18 | 0.4272 | 62.45 |
| 70.0 | 0.03118 | 663.1 | 22.07 | 404.1 | 11.26 | 0.4132 | 56.80 |
| 72.5 | 0.03470 | 665.0 | 22.30 | 390.6 | 11.34 | 0.4001 | 51.75 |
| 75.0 | 0.03856 | 666.8 | 22.53 | 377.9 | 11.42 | 0.3876 | 47.22 |
| 77.5 | 0.04278 | 668.4 | 22.77 | 365.9 | 11.51 | 0.3759 | 43.16 |
| 80.0 | 0.04737 | 670.0 | 23.01 | 354.5 | 11.59 | 0.3648 | 39.51 |
| 82.5 | 0.05238 | 671.5 | 23.25 | 343.7 | 11.68 | 0.3543 | 36.23 |
| 85.0 | 0.05781 | 672.8 | 23.50 | 333.5 | 11.76 | 0.3443 | 33.27 |
| 87.5 | 0.06372 | 674.1 | 23.76 | 323.8 | 11.84 | 0.3348 | 30.59 |
| 90.0 | 0.07012 | 675.3 | 24.02 | 314.5 | 11.93 | 0.3258 | 28.17 |
| 92.5 | 0.07704 | 676.4 | 24.28 | 305.8 | 12.01 | 0.3173 | 25.98 |
| 95.0 | 0.08453 | 677.4 | 24.55 | 297.4 | 12.10 | 0.3092 | 23.99 |
| 97.5 | 0.09261 | 678.3 | 24.82 | 289.5 | 12.18 | 0.3015 | 22.18 |
| 99.63 | 0.10000 | 679.0 | 25.05 | 283.0 | 12.26 | 0.2952 | 20.77 |
| 100.0 | 0.10130 | 679.1 | 25.09 | 281.9 | 12.27 | 0.2941 | 20.53 |

| $a'$ | $a''$ | $Pr'$ | $Pr''$ | $\sigma$ | $b$ | $\beta_v'$ | $\beta_s'$ |
|------|-------|-------|--------|----------|-----|------------|------------|
| $10^{-6}$ m²/s | | — | | mN/m | mm | 1/K | |
| 0.1478 | 328.3 | 5.422 | 1.003 | 71.20 | 2.700 | 160.4 | 10611 |
| 0.1488 | 289.3 | 5.113 | 1.003 | 70.80 | 2.694 | 149.5 | 8524 |
| 0.1499 | 255.6 | 4.830 | 1.002 | 70.41 | 2.688 | 138.7 | 6912 |
| 0.1509 | 226.4 | 4.571 | 1.002 | 70.01 | 2.681 | 128.1 | 5651 |
| 0.1519 | 200.9 | 4.333 | 1.002 | 69.60 | 2.675 | 118.1 | 4652 |
| 0.1529 | 178.8 | 4.114 | 1.001 | 69.19 | 2.668 | 108.6 | 3855 |
| 0.1539 | 159.5 | 3.913 | 1.001 | 68.78 | 2.662 | 99.66 | 3212 |
| 0.1548 | 142.6 | 3.727 | 1.001 | 68.37 | 2.655 | 91.38 | 2691 |
| 0.1558 | 127.7 | 3.555 | 1.001 | 67.95 | 2.648 | 83.72 | 2265 |
| 0.1566 | 114.6 | 3.396 | 1.000 | 67.53 | 2.642 | 76.65 | 1915 |
| 0.1575 | 103.1 | 3.248 | 1.000 | 67.10 | 2.635 | 70.16 | 1626 |
| 0.1583 | 92.93 | 3.110 | 1.000 | 66.67 | 2.628 | 64.21 | 1386 |
| 0.1591 | 83.92 | 2.983 | 1.000 | 66.24 | 2.621 | 58.76 | 1186 |
| 0.1599 | 75.93 | 2.864 | 1.000 | 65.81 | 2.614 | 53.78 | 1018 |
| 0.1606 | 68.82 | 2.752 | 0.999 | 65.37 | 2.608 | 49.24 | 877.0 |
| 0.1613 | 62.49 | 2.649 | 0.999 | 64.93 | 2.601 | 45.09 | 757.8 |
| 0.1620 | 56.85 | 2.551 | 0.999 | 64.49 | 2.594 | 41.31 | 656.7 |
| 0.1626 | 51.79 | 2.460 | 0.999 | 64.04 | 2.587 | 37.86 | 570.7 |
| 0.1632 | 47.27 | 2.375 | 0.999 | 63.59 | 2.579 | 34.71 | 497.3 |
| 0.1638 | 43.21 | 2.295 | 0.999 | 63.13 | 2.572 | 31.85 | 434.5 |
| 0.1644 | 39.56 | 2.219 | 0.999 | 62.68 | 2.565 | 29.23 | 380.5 |
| 0.1649 | 36.27 | 2.148 | 0.999 | 62.22 | 2.558 | 26.85 | 334.0 |
| 0.1654 | 33.31 | 2.081 | 0.999 | 61.76 | 2.550 | 24.68 | 293.9 |
| 0.1659 | 30.63 | 2.018 | 0.999 | 61.29 | 2.543 | 22.70 | 259.2 |
| 0.1664 | 28.20 | 1.958 | 0.999 | 60.82 | 2.535 | 20.89 | 229.1 |
| 0.1668 | 26.01 | 1.902 | 0.999 | 60.35 | 2.528 | 19.23 | 202.9 |
| 0.1672 | 24.01 | 1.849 | 0.999 | 59.88 | 2.520 | 17.72 | 180.0 |
| 0.1676 | 22.20 | 1.798 | 0.999 | 59.40 | 2.512 | 16.34 | 160.0 |
| 0.1680 | 20.78 | 1.757 | 0.999 | 58.99 | 2.506 | 15.26 | 144.9 |
| 0.1680 | 20.55 | 1.750 | 0.999 | 58.92 | 2.505 | 15.08 | 142.5 |

Table 2, part 2 (cont.)

| $t$ | $p$ | $\lambda'$ | $\lambda''$ | $\eta'$ | $\eta''$ | $\nu'$ | $\nu''$ |
|---|---|---|---|---|---|---|---|
| °C | MPa | mW/Km | | μPa s | | $10^{-6}$ m²/s | |
| 105.0 | 0.1208 | 680.6 | 25.66 | 267.7 | 12.44 | 0.2804 | 17.66 |
| 110.0 | 0.1432 | 681.7 | 26.24 | 254.8 | 12.61 | 0.2680 | 15.27 |
| 115.0 | 0.1690 | 682.6 | 26.84 | 243.0 | 12.78 | 0.2566 | 13.26 |
| 120.0 | 0.1985 | 683.2 | 27.46 | 232.2 | 12.96 | 0.2462 | 11.56 |
| 125.0 | 0.2320 | 683.6 | 28.10 | 222.2 | 13.13 | 0.2366 | 10.12 |
| 130.0 | 0.2700 | 683.7 | 28.76 | 213.0 | 13.30 | 0.2278 | 8.894 |
| 135.0 | 0.3129 | 683.6 | 29.44 | 204.5 | 13.47 | 0.2198 | 7.846 |
| 140.0 | 0.3612 | 683.3 | 30.14 | 196.6 | 13.65 | 0.2123 | 6.946 |
| 145.0 | 0.4153 | 682.8 | 30.85 | 189.3 | 13.82 | 0.2054 | 6.169 |
| 150.0 | 0.4757 | 682.1 | 31.59 | 182.5 | 13.99 | 0.1991 | 5.496 |
| 155.0 | 0.5430 | 681.1 | 32.35 | 176.2 | 14.16 | 0.1932 | 4.912 |
| 160.0 | 0.6177 | 680.0 | 33.12 | 170.3 | 14.34 | 0.1877 | 4.402 |
| 165.0 | 0.7003 | 678.6 | 33.92 | 164.8 | 14.51 | 0.1826 | 3.956 |
| 170.0 | 0.7915 | 677.1 | 34.74 | 159.6 | 14.68 | 0.1779 | 3.565 |
| 175.0 | 0.8918 | 675.3 | 35.58 | 154.8 | 14.85 | 0.1734 | 3.220 |
| 180.0 | 1.0019 | 673.4 | 36.44 | 150.2 | 15.02 | 0.1693 | 2.915 |
| 185.0 | 1.1225 | 671.2 | 37.32 | 145.9 | 15.20 | 0.1655 | 2.645 |
| 190.0 | 1.2542 | 668.8 | 38.23 | 141.8 | 15.37 | 0.1619 | 2.405 |
| 195.0 | 1.3976 | 666.2 | 39.15 | 138.0 | 15.54 | 0.1585 | 2.192 |
| 200.0 | 1.5536 | 663.4 | 40.10 | 134.4 | 15.71 | 0.1554 | 2.001 |
| 205.0 | 1.7229 | 660.3 | 41.08 | 130.9 | 15.89 | 0.1525 | 1.830 |
| 210.0 | 1.9062 | 657.1 | 42.07 | 127.7 | 16.06 | 0.1497 | 1.676 |
| 215.0 | 2.1042 | 653.5 | 43.10 | 124.5 | 16.23 | 0.1471 | 1.538 |
| 220.0 | 2.3178 | 649.8 | 44.15 | 121.6 | 16.41 | 0.1447 | 1.414 |
| 225.0 | 2.5479 | 645.7 | 45.24 | 118.7 | 16.59 | 0.1424 | 1.301 |
| 230.0 | 2.7951 | 641.4 | 46.35 | 116.0 | 16.76 | 0.1403 | 1.199 |
| 235.0 | 3.0604 | 636.9 | 47.51 | 113.4 | 16.94 | 0.1382 | 1.107 |
| 240.0 | 3.3447 | 632.0 | 48.70 | 110.9 | 17.12 | 0.1363 | 1.023 |
| 245.0 | 3.6488 | 626.8 | 49.94 | 108.5 | 17.31 | 0.1346 | 0.946 |
| 250.0 | 3.9736 | 621.4 | 51.23 | 106.2 | 17.49 | 0.1329 | 0.877 |

| $a'$ | $a''$ | $Pr'$ | $Pr''$ | $\sigma$ | $b$ | $\beta_v'$ | $\beta_s'$ |
|---|---|---|---|---|---|---|---|
| $10^{-6}$ m²/s | | — | | mN/m | mm | 1/K | |
| 0.1687 | 17.67 | 1.662 | 1.000 | 57.95 | 2.489 | 12.87 | 113.6 |
| 0.1694 | 15.26 | 1.582 | 1.001 | 56.97 | 2.473 | 11.01 | 91.23 |
| 0.1700 | 13.24 | 1.509 | 1.001 | 55.98 | 2.456 | 9.442 | 73.69 |
| 0.1705 | 11.53 | 1.444 | 1.003 | 54.97 | 2.439 | 8.122 | 59.88 |
| 0.1710 | 10.08 | 1.384 | 1.004 | 53.96 | 2.422 | 7.005 | 48.92 |
| 0.1714 | 8.840 | 1.329 | 1.006 | 52.94 | 2.405 | 6.057 | 40.17 |
| 0.1717 | 7.781 | 1.280 | 1.008 | 51.91 | 2.387 | 5.251 | 33.15 |
| 0.1720 | 6.869 | 1.234 | 1.011 | 50.86 | 2.369 | 4.563 | 27.49 |
| 0.1723 | 6.082 | 1.192 | 1.014 | 49.81 | 2.350 | 3.976 | 22.89 |
| 0.1725 | 5.399 | 1.154 | 1.018 | 48.75 | 2.331 | 3.471 | 19.14 |
| 0.1726 | 4.804 | 1.119 | 1.022 | 47.68 | 2.312 | 3.038 | 16.07 |
| 0.1727 | 4.285 | 1.087 | 1.027 | 46.60 | 2.292 | 2.665 | 13.53 |
| 0.1727 | 3.830 | 1.057 | 1.033 | 45.51 | 2.272 | 2.342 | 11.44 |
| 0.1727 | 3.430 | 1.030 | 1.039 | 44.41 | 2.252 | 2.063 | 9.705 |
| 0.1726 | 3.077 | 1.005 | 1.047 | 43.31 | 2.230 | 1.821 | 8.257 |
| 0.1724 | 2.764 | 0.982 | 1.055 | 42.20 | 2.209 | 1.610 | 7.046 |
| 0.1721 | 2.487 | 0.961 | 1.063 | 41.08 | 2.187 | 1.426 | 6.029 |
| 0.1718 | 2.241 | 0.942 | 1.073 | 39.95 | 2.164 | 1.265 | 5.172 |
| 0.1714 | 2.021 | 0.925 | 1.084 | 38.82 | 2.141 | 1.124 | 4.448 |
| 0.1709 | 1.825 | 0.909 | 1.096 | 37.68 | 2.118 | 1.001 | 3.834 |
| 0.1703 | 1.650 | 0.895 | 1.109 | 36.54 | 2.093 | 0.892 | 3.312 |
| 0.1696 | 1.492 | 0.883 | 1.123 | 35.39 | 2.069 | 0.797 | 2.867 |
| 0.1688 | 1.351 | 0.871 | 1.139 | 34.24 | 2.043 | 0.712 | 2.486 |
| 0.1680 | 1.224 | 0.861 | 1.155 | 33.08 | 2.017 | 0.637 | 2.160 |
| 0.1670 | 1.109 | 0.853 | 1.174 | 31.91 | 1.991 | 0.571 | 1.879 |
| 0.1659 | 1.005 | 0.846 | 1.193 | 30.75 | 1.963 | 0.512 | 1.637 |
| 0.1646 | 0.912 | 0.840 | 1.214 | 29.58 | 1.935 | 0.460 | 1.429 |
| 0.1633 | 0.827 | 0.835 | 1.237 | 28.40 | 1.907 | 0.413 | 1.248 |
| 0.1618 | 0.750 | 0.832 | 1.262 | 27.23 | 1.877 | 0.372 | 1.091 |
| 0.1601 | 0.680 | 0.830 | 1.288 | 26.05 | 1.847 | 0.335 | 0.955 |

Table 2, part 2 (cont.)

| t | p | λ' | λ" | η' | η" | ν' | ν" |
|---|---|----|----|----|----|----|----|
| °C | MPa | mW/K m | | μPa s | | $10^{-6}$ m²/s | |
| 255.0 | 4.3202 | 615.6 | 52.57 | 103.9 | 17.68 | 0.1313 | 0.813 |
| 260.0 | 4.6894 | 609.4 | 53.98 | 101.7 | 17.88 | 0.1298 | 0.754 |
| 265.0 | 5.0823 | 603.0 | 55.47 | 99.62 | 18.07 | 0.1284 | 0.701 |
| 270.0 | 5.4999 | 596.1 | 57.04 | 97.56 | 18.27 | 0.1271 | 0.651 |
| 275.0 | 5.9431 | 588.9 | 58.72 | 95.54 | 18.48 | 0.1258 | 0.606 |
| 280.0 | 6.4132 | 581.4 | 60.52 | 93.57 | 18.70 | 0.1247 | 0.564 |
| 285.0 | 6.9111 | 573.5 | 62.47 | 91.63 | 18.92 | 0.1236 | 0.525 |
| 290.0 | 7.4380 | 565.2 | 64.59 | 89.72 | 19.15 | 0.1225 | 0.490 |
| 295.0 | 7.9952 | 556.6 | 66.91 | 87.83 | 19.39 | 0.1216 | 0.456 |
| 300.0 | 8.5838 | 547.7 | 69.49 | 85.96 | 19.65 | 0.1207 | 0.426 |
| 305.0 | 9.2051 | 538.4 | 72.36 | 84.09 | 19.92 | 0.1198 | 0.397 |
| 310.0 | 9.8605 | 529.0 | 75.61 | 82.22 | 20.20 | 0.1190 | 0.371 |
| 315.0 | 10.551 | 519.3 | 79.32 | 80.35 | 20.51 | 0.1183 | 0.346 |
| 320.0 | 11.279 | 509.4 | 83.59 | 78.46 | 20.84 | 0.1176 | 0.323 |
| 325.0 | 12.046 | 499.3 | 88.58 | 76.54 | 21.21 | 0.1169 | 0.301 |
| 330.0 | 12.852 | 489.2 | 94.48 | 74.58 | 21.60 | 0.1163 | 0.281 |
| 335.0 | 13.701 | 478.9 | 101.6 | 72.55 | 22.05 | 0.1158 | 0.261 |
| 340.0 | 14.594 | 468.6 | 110.2 | 70.45 | 22.55 | 0.1153 | 0.243 |
| 345.0 | 15.533 | 458.1 | 120.9 | 68.24 | 23.13 | 0.1150 | 0.226 |
| 350.0 | 16.521 | 447.6 | 134.6 | 65.88 | 23.81 | 0.1146 | 0.210 |
| 355.0 | 17.561 | 437.1 | 152.8 | 63.30 | 24.65 | 0.1144 | 0.194 |
| 360.0 | 18.655 | 427.2 | 178.0 | 60.39 | 25.71 | 0.1144 | 0.179 |
| 365.0 | 19.809 | 419.9 | 216.9 | 56.93 | 27.18 | 0.1145 | 0.164 |
| 370.0 | 21.030 | 428.0 | 299.4 | 52.26 | 29.57 | 0.1153 | 0.148 |
| 371.0 | 21.283 | 438.4 | 334.9 | 50.97 | 30.33 | 0.1156 | 0.144 |
| 372.0 | 21.539 | 459.9 | 400.7 | 49.44 | 31.36 | 0.1161 | 0.140 |
| 373.0 | 21.799 | 531.1 | 565.1 | 48.02 | 33.43 | 0.1189 | 0.137 |
| 374.0 | 22.055 | 1419 | | 43.2 | | 0.134 | |

| $a'$ | $a''$ | $Pr'$ | $Pr''$ | $\sigma$ | $b$ | $\beta_v'$ | $\beta_s'$ |
|---|---|---|---|---|---|---|---|
| $10^{-6}$ m²/s | | — | | mN/m | mm | 1/K | |
| 0.1583 | 0.617 | 0.829 | 1.317 | 24.88 | 1.815 | 0.302 | 0.836 |
| 0.1564 | 0.560 | 0.830 | 1.347 | 23.70 | 1.783 | 0.272 | 0.733 |
| 0.1542 | 0.508 | 0.833 | 1.380 | 22.52 | 1.750 | 0.245 | 0.643 |
| 0.1519 | 0.460 | 0.836 | 1.415 | 21.35 | 1.716 | 0.221 | 0.563 |
| 0.1494 | 0.417 | 0.842 | 1.453 | 20.17 | 1.680 | 0.199 | 0.494 |
| 0.1467 | 0.378 | 0.850 | 1.494 | 19.00 | 1.644 | 0.180 | 0.433 |
| 0.1438 | 0.342 | 0.859 | 1.538 | 17.84 | 1.606 | 0.162 | 0.380 |
| 0.1407 | 0.309 | 0.871 | 1.585 | 16.68 | 1.566 | 0.146 | 0.332 |
| 0.1374 | 0.279 | 0.885 | 1.636 | 15.52 | 1.526 | 0.132 | 0.291 |
| 0.1338 | 0.252 | 0.902 | 1.691 | 14.37 | 1.483 | 0.119 | 0.254 |
| 0.1300 | 0.227 | 0.922 | 1.751 | 13.23 | 1.439 | 0.107 | 0.222 |
| 0.1258 | 0.204 | 0.946 | 1.817 | 12.10 | 1.392 | 0.096 | 0.193 |
| 0.1214 | 0.183 | 0.974 | 1.889 | 10.98 | 1.344 | 0.086 | 0.168 |
| 0.1167 | 0.164 | 1.008 | 1.969 | 9.875 | 1.293 | 0.077 | 0.145 |
| 0.1115 | 0.146 | 1.048 | 2.059 | 8.785 | 1.238 | 0.069 | 0.125 |
| 0.1060 | 0.130 | 1.098 | 2.163 | 7.713 | 1.181 | 0.062 | 0.108 |
| 0.0999 | 0.114 | 1.160 | 2.283 | 6.662 | 1.119 | 0.055 | 0.092 |
| 0.0931 | 0.100 | 1.239 | 2.428 | 5.636 | 1.053 | 0.049 | 0.078 |
| 0.0856 | 0.087 | 1.344 | 2.610 | 4.638 | 0.9811 | 0.043 | 0.066 |
| 0.0769 | 0.074 | 1.490 | 2.849 | 3.675 | 0.9014 | 0.038 | 0.055 |
| 0.0669 | 0.061 | 1.710 | 3.189 | 2.754 | 0.8117 | 0.033 | 0.045 |
| 0.0551 | 0.048 | 2.077 | 3.726 | 1.886 | 0.7073 | 0.028 | 0.036 |
| 0.0408 | 0.035 | 2.810 | 4.735 | 1.088 | 0.5789 | 0.024 | 0.028 |
| 0.0225 | 0.019 | 5.122 | 7.780 | 0.3948 | 0.3991 | 0.019 | 0.021 |
| 0.0179 | 0.015 | 6.463 | 9.455 | 0.2754 | 0.3493 | 0.018 | 0.019 |
| 0.0128 | 0.011 | 9.037 | 13.04 | 0.1656 | 0.2897 | 0.017 | 0.017 |
| 0.0067 | 0.006 | 17.81 | 24.30 | 0.0694 | 0.2109 | 0.015 | 0.015 |
| 0 | | $\infty$ | | 0 | 0 | 0.01214 | |

93

Table 2, part 3

| $t$ | $p$ | $\alpha_p'$ | $\alpha_p''$ | $\delta_h'$ | $\delta_h''$ | $\delta_T'$ | $\delta_T''$ | $\frac{dr}{dT}$ |
|---|---|---|---|---|---|---|---|---|
| °C | MPa | $10^{-3}$/K | | K/MPa | | $10^{-3}$m³/kg | | kJ/kg K |
| 0.0[1] | 0.00061 | −0.0806 | 3.672 | −0.2417 | 331.0 | 1.022 | −618.2 | −2.389 |
| 1.0 | 0.00066 | −0.0608 | 3.659 | −0.2409 | 323.6 | 1.017 | −604.5 | −2.382 |
| 2.0 | 0.00071 | −0.0417 | 3.646 | −0.2400 | 316.4 | 1.012 | −591.2 | −2.376 |
| 3.0 | 0.00076 | −0.0235 | 3.634 | −0.2391 | 309.4 | 1.007 | −578.3 | −2.371 |
| 4.0 | 0.00081 | −0.0059 | 3.621 | −0.2383 | 302.6 | 1.002 | −565.7 | −2.367 |
| 5.0 | 0.00087 | 0.0110 | 3.609 | −0.2374 | 295.9 | 0.9970 | −553.5 | −2.363 |
| 6.0 | 0.00094 | 0.0273 | 3.596 | −0.2365 | 289.5 | 0.9924 | −541.7 | −2.361 |
| 7.0 | 0.00100 | 0.0431 | 3.584 | −0.2356 | 283.2 | 0.9880 | −530.2 | −2.359 |
| 8.0 | 0.00107 | 0.0583 | 3.572 | −0.2347 | 277.1 | 0.9838 | −518.9 | −2.357 |
| 9.0 | 0.00115 | 0.0730 | 3.560 | −0.2338 | 271.2 | 0.9796 | −508.1 | −2.356 |
| 10.0 | 0.00123 | 0.0872 | 3.548 | −0.2329 | 265.5 | 0.9756 | −497.5 | −2.355 |
| 11.0 | 0.00131 | 0.1010 | 3.536 | −0.2321 | 259.9 | 0.9717 | −487.2 | −2.355 |
| 12.0 | 0.00140 | 0.1143 | 3.525 | −0.2312 | 254.4 | 0.9679 | −477.2 | −2.355 |
| 13.0 | 0.00150 | 0.1273 | 3.513 | −0.2304 | 249.1 | 0.9642 | −467.4 | −2.355 |
| 14.0 | 0.00160 | 0.1399 | 3.502 | −0.2295 | 244.0 | 0.9606 | −457.9 | −2.355 |
| 15.0 | 0.00171 | 0.1522 | 3.490 | −0.2287 | 239.0 | 0.9570 | −448.7 | −2.356 |
| 16.0 | 0.00182 | 0.1641 | 3.479 | −0.2279 | 234.1 | 0.9536 | −439.8 | −2.357 |
| 17.0 | 0.00194 | 0.1757 | 3.468 | −0.2271 | 229.4 | 0.9502 | −431.0 | −2.357 |
| 18.0 | 0.00206 | 0.1870 | 3.457 | −0.2263 | 224.7 | 0.9469 | −422.5 | −2.358 |
| 19.0 | 0.00220 | 0.1981 | 3.446 | −0.2256 | 220.2 | 0.9436 | −414.3 | −2.359 |
| 20.0 | 0.00234 | 0.2089 | 3.435 | −0.2248 | 215.9 | 0.9405 | −406.2 | −2.360 |
| 21.0 | 0.00249 | 0.2194 | 3.424 | −0.2241 | 211.6 | 0.9374 | −398.4 | −2.361 |
| 22.0 | 0.00264 | 0.2297 | 3.414 | −0.2233 | 207.4 | 0.9343 | −390.8 | −2.363 |
| 23.0 | 0.00281 | 0.2398 | 3.403 | −0.2226 | 203.4 | 0.9313 | −383.3 | −2.364 |
| 24.0 | 0.00299 | 0.2497 | 3.393 | −0.2219 | 199.4 | 0.9283 | −376.1 | −2.365 |
| 25.0 | 0.00317 | 0.2593 | 3.383 | −0.2212 | 195.6 | 0.9254 | −369.1 | −2.366 |
| 26.0 | 0.00336 | 0.2688 | 3.372 | −0.2205 | 191.9 | 0.9226 | −362.2 | −2.368 |
| 27.0 | 0.00357 | 0.2781 | 3.362 | −0.2199 | 188.2 | 0.9198 | −355.5 | −2.369 |
| 28.0 | 0.00378 | 0.2872 | 3.352 | −0.2192 | 184.7 | 0.9170 | −349.0 | −2.371 |
| 29.0 | 0.00401 | 0.2962 | 3.342 | −0.2185 | 181.2 | 0.9142 | −342.6 | −2.372 |

[1] The values of this line relate to metastable states.

| $\dfrac{dp}{dT}\Big|_{sat}$ | $\dfrac{d^2p}{dT^2}\Big|_{sat}$ | $c'_{sat}$ | $c''_{sat}$ | $c'$ | $c''$ | $\chi'_T$ | $\chi''_T$ |
|---|---|---|---|---|---|---|---|
| kPa/K | Pa/K² | kJ/kg K | | km/s | | 1/GPa | |
| 0.0444 | 2.860 | 4.229 | −7.314 | 1.401 | 0.4093 | 0.5101 | 1636700 |
| 0.0474 | 3.023 | 4.221 | −7.273 | 1.406 | 0.4101 | 0.5061 | 1522500 |
| 0.0505 | 3.193 | 4.215 | −7.232 | 1.411 | 0.4108 | 0.5022 | 1417100 |
| 0.0537 | 3.371 | 4.209 | −7.191 | 1.416 | 0.4115 | 0.4986 | 1319800 |
| 0.0572 | 3.556 | 4.204 | −7.151 | 1.421 | 0.4123 | 0.4951 | 1229800 |
| 0.0609 | 3.751 | 4.200 | −7.111 | 1.426 | 0.4130 | 0.4918 | 1146700 |
| 0.0647 | 3.953 | 4.197 | −7.071 | 1.431 | 0.4137 | 0.4887 | 1069800 |
| 0.0688 | 4.165 | 4.194 | −7.031 | 1.435 | 0.4144 | 0.4857 | 998590 |
| 0.0730 | 4.386 | 4.192 | −6.992 | 1.439 | 0.4152 | 0.4829 | 932660 |
| 0.0775 | 4.616 | 4.190 | −6.953 | 1.444 | 0.4159 | 0.4803 | 871540 |
| 0.0823 | 4.855 | 4.188 | −6.915 | 1.448 | 0.4166 | 0.4777 | 814880 |
| 0.0873 | 5.105 | 4.187 | −6.877 | 1.452 | 0.4173 | 0.4753 | 762300 |
| 0.0925 | 5.365 | 4.186 | −6.839 | 1.456 | 0.4180 | 0.4731 | 713500 |
| 0.0980 | 5.636 | 4.185 | −6.801 | 1.459 | 0.4187 | 0.4709 | 668180 |
| 0.1038 | 5.917 | 4.185 | −6.764 | 1.463 | 0.4195 | 0.4688 | 626050 |
| 0.1098 | 6.209 | 4.184 | −6.727 | 1.467 | 0.4202 | 0.4669 | 586890 |
| 0.1162 | 6.514 | 4.184 | −6.690 | 1.470 | 0.4209 | 0.4651 | 550450 |
| 0.1229 | 6.829 | 4.184 | −6.654 | 1.473 | 0.4216 | 0.4633 | 516530 |
| 0.1298 | 7.157 | 4.183 | −6.617 | 1.477 | 0.4223 | 0.4616 | 484950 |
| 0.1372 | 7.498 | 4.183 | −6.581 | 1.480 | 0.4230 | 0.4601 | 455520 |
| 0.1448 | 7.851 | 4.183 | −6.546 | 1.483 | 0.4237 | 0.4586 | 428080 |
| 0.1529 | 8.217 | 4.183 | −6.511 | 1.486 | 0.4244 | 0.4572 | 402490 |
| 0.1613 | 8.596 | 4.183 | −6.476 | 1.489 | 0.4251 | 0.4558 | 378610 |
| 0.1701 | 8.990 | 4.183 | −6.441 | 1.492 | 0.4258 | 0.4546 | 356310 |
| 0.1793 | 9.397 | 4.183 | −6.406 | 1.494 | 0.4265 | 0.4534 | 335480 |
| 0.1889 | 9.819 | 4.183 | −6.372 | 1.497 | 0.4271 | 0.4523 | 316020 |
| 0.1989 | 10.26 | 4.183 | −6.338 | 1.500 | 0.4278 | 0.4513 | 297820 |
| 0.2094 | 10.71 | 4.183 | −6.304 | 1.502 | 0.4285 | 0.4503 | 280800 |
| 0.2203 | 11.17 | 4.183 | −6.271 | 1.505 | 0.4292 | 0.4494 | 264860 |
| 0.2317 | 11.66 | 4.183 | −6.238 | 1.507 | 0.4299 | 0.4485 | 249950 |

(1) Die Werte dieser Zeile gelten für metastabile Zustände.

| t | p | $\alpha_p'$ | $\alpha_p''$ | $\delta_h'$ | $\delta_h''$ | $\delta_T'$ | $\delta_T''$ | $\frac{dr}{dT}$ |
|---|---|---|---|---|---|---|---|---|
| °C | MPa | $10^{-3}$/K | | K/MPa | | $10^{-3}$m$^3$/kg | | kJ/kg K |
| 30.0 | 0.00425 | 0.3050 | 3.332 | −0.2179 | 177.8 | 0.9115 | −336.5 | −2.373 |
| 32.5 | 0.00489 | 0.3264 | 3.308 | −0.2163 | 169.7 | 0.9049 | −321.7 | −2.377 |
| 35.0 | 0.00563 | 0.3469 | 3.285 | −0.2148 | 162.2 | 0.8985 | −307.8 | −2.381 |
| 37.5 | 0.00645 | 0.3667 | 3.262 | −0.2133 | 155.0 | 0.8922 | −294.7 | −2.385 |
| 40.0 | 0.00738 | 0.3859 | 3.240 | −0.2119 | 148.3 | 0.8861 | −282.5 | −2.389 |
| 42.5 | 0.00842 | 0.4045 | 3.218 | −0.2104 | 142.0 | 0.8801 | −270.9 | −2.394 |
| 45.0 | 0.00959 | 0.4225 | 3.197 | −0.2090 | 136.1 | 0.8742 | −260.1 | −2.399 |
| 47.5 | 0.01089 | 0.4401 | 3.176 | −0.2077 | 130.5 | 0.8684 | −249.9 | −2.404 |
| 50.0 | 0.01234 | 0.4572 | 3.156 | −0.2063 | 125.2 | 0.8626 | −240.3 | −2.410 |
| 52.5 | 0.01396 | 0.4740 | 3.137 | −0.2049 | 120.2 | 0.8569 | −231.2 | −2.416 |
| 55.0 | 0.01575 | 0.4903 | 3.118 | −0.2036 | 115.5 | 0.8513 | −222.6 | −2.423 |
| 57.5 | 0.01774 | 0.5064 | 3.100 | −0.2022 | 111.0 | 0.8457 | −214.6 | −2.430 |
| 60.0 | 0.01993 | 0.5222 | 3.083 | −0.2009 | 106.8 | 0.8402 | −206.9 | −2.437 |
| 62.5 | 0.02235 | 0.5377 | 3.066 | −0.1995 | 102.8 | 0.8347 | −199.7 | −2.445 |
| 65.0 | 0.02502 | 0.5529 | 3.049 | −0.1982 | 99.07 | 0.8292 | −192.9 | −2.454 |
| 67.5 | 0.02796 | 0.5679 | 3.034 | −0.1968 | 95.49 | 0.8237 | −186.5 | −2.463 |
| 70.0 | 0.03118 | 0.5827 | 3.018 | −0.1954 | 92.10 | 0.8182 | −180.4 | −2.473 |
| 72.5 | 0.03470 | 0.5974 | 3.004 | −0.1941 | 88.89 | 0.8128 | −174.6 | −2.483 |
| 75.0 | 0.03856 | 0.6118 | 2.990 | −0.1927 | 85.84 | 0.8073 | −169.1 | −2.494 |
| 77.5 | 0.04278 | 0.6261 | 2.976 | −0.1913 | 82.94 | 0.8018 | −163.9 | −2.506 |
| 80.0 | 0.04737 | 0.6403 | 2.964 | −0.1899 | 80.18 | 0.7963 | −159.0 | −2.518 |
| 82.5 | 0.05238 | 0.6543 | 2.952 | −0.1885 | 77.56 | 0.7908 | −154.3 | −2.531 |
| 85.0 | 0.05781 | 0.6683 | 2.940 | −0.1870 | 75.07 | 0.7853 | −149.9 | −2.544 |
| 87.5 | 0.06372 | 0.6821 | 2.929 | −0.1856 | 72.69 | 0.7797 | −145.7 | −2.558 |
| 90.0 | 0.07012 | 0.6958 | 2.919 | −0.1841 | 70.43 | 0.7742 | −141.6 | −2.572 |
| 92.5 | 0.07704 | 0.7095 | 2.909 | −0.1827 | 68.27 | 0.7685 | −137.8 | −2.588 |
| 95.0 | 0.08453 | 0.7230 | 2.900 | −0.1812 | 66.21 | 0.7629 | −134.2 | −2.603 |
| 97.5 | 0.09261 | 0.7366 | 2.892 | −0.1797 | 64.24 | 0.7571 | −130.8 | −2.620 |
| 99.63 | 0.10000 | 0.7481 | 2.885 | −0.1784 | 62.63 | 0.7522 | −127.9 | −2.634 |
| 100.0 | 0.10132 | 0.7501 | 2.884 | −0.1782 | 62.36 | 0.7514 | −127.5 | −2.637 |

| $\dfrac{dp}{dT}\bigg|_{\text{sat}}$ | $\dfrac{d^2p}{dT^2}\bigg|_{\text{sat}}$ | $c'_{\text{sat}}$ | $c''_{\text{sat}}$ | $c'$ | $c''$ | $\chi'_T$ | $\chi''_T$ |
|---|---|---|---|---|---|---|---|
| kPa/K | Pa/K² | kJ/kg K | | km/s | | 1/GPa | |
| 0.2436 | 12.16 | 4.183 | −6.205 | 1.509 | 0.4306 | 0.4477 | 235970 |
| 0.2757 | 13.47 | 4.183 | −6.124 | 1.515 | 0.4323 | 0.4460 | 204740 |
| 0.3111 | 14.90 | 4.183 | −6.044 | 1.520 | 0.4339 | 0.4446 | 178120 |
| 0.3503 | 16.44 | 4.183 | −5.966 | 1.524 | 0.4356 | 0.4436 | 155350 |
| 0.3934 | 18.11 | 4.182 | −5.890 | 1.528 | 0.4373 | 0.4428 | 135830 |
| 0.4409 | 19.89 | 4.182 | −5.815 | 1.532 | 0.4389 | 0.4423 | 119060 |
| 0.4930 | 21.81 | 4.182 | −5.742 | 1.535 | 0.4405 | 0.4421 | 104610 |
| 0.5501 | 23.86 | 4.182 | −5.670 | 1.539 | 0.4422 | 0.4421 | 92124 |
| 0.6124 | 26.05 | 4.182 | −5.599 | 1.541 | 0.4438 | 0.4424 | 81313 |
| 0.6804 | 28.39 | 4.182 | −5.530 | 1.544 | 0.4453 | 0.4429 | 71930 |
| 0.7545 | 30.88 | 4.182 | −5.462 | 1.546 | 0.4469 | 0.4437 | 63768 |
| 0.8350 | 33.53 | 4.182 | −5.396 | 1.548 | 0.4485 | 0.4446 | 56651 |
| 0.9223 | 36.34 | 4.183 | −5.331 | 1.549 | 0.4500 | 0.4458 | 50433 |
| 1.017 | 39.31 | 4.183 | −5.268 | 1.551 | 0.4516 | 0.4472 | 44987 |
| 1.119 | 42.46 | 4.184 | −5.206 | 1.552 | 0.4531 | 0.4488 | 40209 |
| 1.229 | 45.78 | 4.185 | −5.145 | 1.552 | 0.4546 | 0.4506 | 36007 |
| 1.348 | 49.29 | 4.187 | −5.085 | 1.553 | 0.4561 | 0.4525 | 32305 |
| 1.476 | 52.98 | 4.188 | −5.027 | 1.553 | 0.4576 | 0.4547 | 29036 |
| 1.613 | 56.86 | 4.190 | −4.970 | 1.553 | 0.4590 | 0.4571 | 26145 |
| 1.760 | 60.93 | 4.192 | −4.915 | 1.553 | 0.4605 | 0.4596 | 23584 |
| 1.918 | 65.20 | 4.194 | −4.860 | 1.553 | 0.4619 | 0.4623 | 21310 |
| 2.086 | 69.67 | 4.196 | −4.807 | 1.552 | 0.4633 | 0.4653 | 19287 |
| 2.266 | 74.35 | 4.198 | −4.755 | 1.551 | 0.4647 | 0.4684 | 17485 |
| 2.458 | 79.23 | 4.201 | −4.704 | 1.550 | 0.4661 | 0.4717 | 15877 |
| 2.663 | 84.32 | 4.204 | −4.655 | 1.549 | 0.4675 | 0.4751 | 14440 |
| 2.880 | 89.63 | 4.207 | −4.606 | 1.547 | 0.4688 | 0.4788 | 13153 |
| 3.111 | 95.15 | 4.210 | −4.559 | 1.546 | 0.4702 | 0.4827 | 11999 |
| 3.356 | 100.9 | 4.213 | −4.513 | 1.544 | 0.4715 | 0.4867 | 10962 |
| 3.577 | 106.0 | 4.216 | −4.475 | 1.542 | 0.4726 | 0.4903 | 10160 |
| 3.616 | 106.8 | 4.216 | −4.468 | 1.542 | 0.4728 | 0.4909 | 10029 |

Table 2, part 3 (cont.)

| $t$ | $p$ | $\alpha_p'$ | $\alpha_p''$ | $\delta_h'$ | $\delta_h''$ | $\delta_T'$ | $\delta_T''$ | $\frac{dr}{dT}$ |
|------|------|------|------|------|------|------|------|------|
| °C | MPa | $10^{-3}$/K | | K/MPa | | $10^{-3}$m³/kg | | kJ/kg K |
| 105.0 | 0.12079 | 0.7770 | 2.870 | −0.1751 | 58.85 | 0.7397 | −121.4 | −2.672 |
| 110.0 | 0.14324 | 0.8038 | 2.860 | −0.1720 | 55.62 | 0.7277 | −115.8 | −2.711 |
| 115.0 | 0.16902 | 0.8307 | 2.851 | −0.1687 | 52.66 | 0.7154 | −110.7 | −2.751 |
| 120.0 | 0.19848 | 0.8576 | 2.846 | −0.1654 | 49.93 | 0.7028 | −106.1 | −2.795 |
| 125.0 | 0.23201 | 0.8848 | 2.844 | −0.1620 | 47.41 | 0.6897 | −101.9 | −2.841 |
| 130.0 | 0.27002 | 0.9123 | 2.844 | −0.1585 | 45.07 | 0.6763 | −98.06 | −2.889 |
| 135.0 | 0.31293 | 0.9400 | 2.848 | −0.1548 | 42.90 | 0.6623 | −94.53 | −2.941 |
| 140.0 | 0.36119 | 0.9683 | 2.855 | −0.1511 | 40.88 | 0.6478 | −91.29 | −2.995 |
| 145.0 | 0.41529 | 0.9970 | 2.865 | −0.1471 | 39.00 | 0.6327 | −88.32 | −3.052 |
| 150.0 | 0.47572 | 1.026 | 2.878 | −0.1431 | 37.23 | 0.6169 | −85.58 | −3.112 |
| 155.0 | 0.54299 | 1.056 | 2.895 | −0.1388 | 35.57 | 0.6003 | −83.07 | −3.176 |
| 160.0 | 0.61766 | 1.087 | 2.916 | −0.1344 | 34.02 | 0.5830 | −80.75 | −3.242 |
| 165.0 | 0.70029 | 1.119 | 2.940 | −0.1297 | 32.55 | 0.5647 | −78.62 | −3.312 |
| 170.0 | 0.79147 | 1.152 | 2.969 | −0.1249 | 31.16 | 0.5454 | −76.65 | −3.386 |
| 175.0 | 0.89180 | 1.186 | 3.002 | −0.1197 | 29.85 | 0.5251 | −74.84 | −3.463 |
| 180.0 | 1.0019 | 1.221 | 3.039 | −0.1144 | 28.61 | 0.5036 | −73.18 | −3.544 |
| 185.0 | 1.1225 | 1.258 | 3.081 | −0.1087 | 27.43 | 0.4807 | −71.64 | −3.628 |
| 190.0 | 1.2542 | 1.296 | 3.128 | −0.1027 | 26.31 | 0.4564 | −70.23 | −3.717 |
| 195.0 | 1.3976 | 1.336 | 3.180 | −0.0964 | 25.24 | 0.4305 | −68.94 | −3.810 |
| 200.0 | 1.5536 | 1.377 | 3.238 | −0.0897 | 24.22 | 0.4029 | −67.75 | −3.907 |
| 205.0 | 1.7229 | 1.421 | 3.302 | −0.0827 | 23.25 | 0.3732 | −66.67 | −4.009 |
| 210.0 | 1.9062 | 1.467 | 3.372 | −0.0752 | 22.32 | 0.3415 | −65.69 | −4.116 |
| 215.0 | 2.1042 | 1.516 | 3.450 | −0.0672 | 21.44 | 0.3073 | −64.81 | −4.228 |
| 220.0 | 2.3178 | 1.567 | 3.534 | −0.0587 | 20.59 | 0.2704 | −64.02 | −4.345 |
| 225.0 | 2.5479 | 1.622 | 3.627 | −0.0497 | 19.78 | 0.2305 | −63.32 | −4.468 |
| 230.0 | 2.7951 | 1.680 | 3.729 | −0.0401 | 19.00 | 0.1873 | −62.71 | −4.597 |
| 235.0 | 3.0604 | 1.741 | 3.841 | −0.0298 | 18.26 | 0.1403 | −62.18 | −4.733 |
| 240.0 | 3.3447 | 1.808 | 3.963 | −0.0187 | 17.55 | 0.0891 | −61.75 | −4.877 |
| 245.0 | 3.6488 | 1.879 | 4.097 | −0.0069 | 16.87 | 0.0330 | −61.41 | −5.028 |
| 250.0 | 3.9736 | 1.955 | 4.245 | 0.0059 | 16.21 | −0.0285 | −61.16 | −5.187 |

| $\dfrac{dp}{dT}\Big|_{sat}$ | $\dfrac{d^2p}{dT^2}\Big|_{sat}$ | $c'_{sat}$ | $c''_{sat}$ | $c'$ | $c''$ | $\chi'_T$ | $\chi''_T$ |
|---|---|---|---|---|---|---|---|
| kPa/K | Pa/K² | kJ/kg K | | km/s | | 1/GPa | |
| 4.181 | 119.4 | 4.223 | −4.382 | 1.537 | 0.4753 | 0.5000 | 8430 |
| 4.812 | 132.9 | 4.230 | −4.300 | 1.532 | 0.4778 | 0.5098 | 7125 |
| 5.512 | 147.3 | 4.238 | −4.223 | 1.525 | 0.4802 | 0.5205 | 6054 |
| 6.286 | 162.6 | 4.246 | −4.150 | 1.518 | 0.4825 | 0.5320 | 5169 |
| 7.140 | 178.9 | 4.255 | −4.082 | 1.511 | 0.4847 | 0.5444 | 4435 |
| 8.077 | 196.1 | 4.264 | −4.018 | 1.503 | 0.4869 | 0.5578 | 3824 |
| 9.102 | 214.2 | 4.274 | −3.958 | 1.494 | 0.4889 | 0.5721 | 3311 |
| 10.22 | 233.2 | 4.284 | −3.902 | 1.485 | 0.4909 | 0.5874 | 2880 |
| 11.44 | 253.1 | 4.295 | −3.850 | 1.475 | 0.4928 | 0.6039 | 2515 |
| 12.75 | 273.9 | 4.306 | −3.803 | 1.464 | 0.4946 | 0.6215 | 2205 |
| 14.18 | 295.7 | 4.318 | −3.759 | 1.453 | 0.4962 | 0.6404 | 1941 |
| 15.71 | 318.3 | 4.330 | −3.719 | 1.441 | 0.4978 | 0.6606 | 1716 |
| 17.36 | 341.8 | 4.344 | −3.684 | 1.429 | 0.4993 | 0.6823 | 1522 |
| 19.13 | 366.2 | 4.358 | −3.652 | 1.416 | 0.5007 | 0.7055 | 1354 |
| 21.02 | 391.4 | 4.373 | −3.624 | 1.403 | 0.5019 | 0.7304 | 1210 |
| 23.05 | 417.5 | 4.389 | −3.600 | 1.389 | 0.5031 | 0.7571 | 1084 |
| 25.20 | 444.5 | 4.406 | −3.580 | 1.375 | 0.5041 | 0.7858 | 974.9 |
| 27.49 | 472.2 | 4.424 | −3.564 | 1.360 | 0.5050 | 0.8166 | 879.4 |
| 29.92 | 500.8 | 4.444 | −3.551 | 1.345 | 0.5057 | 0.8498 | 795.8 |
| 32.50 | 530.1 | 4.465 | −3.543 | 1.329 | 0.5064 | 0.8856 | 722.4 |
| 35.23 | 560.3 | 4.487 | −3.538 | 1.313 | 0.5069 | 0.9243 | 657.7 |
| 38.11 | 591.2 | 4.510 | −3.538 | 1.296 | 0.5072 | 0.9662 | 600.6 |
| 41.14 | 623.0 | 4.536 | −3.541 | 1.279 | 0.5074 | 1.012 | 550.1 |
| 44.34 | 655.5 | 4.563 | −3.549 | 1.262 | 0.5075 | 1.061 | 505.2 |
| 47.70 | 688.8 | 4.592 | −3.561 | 1.243 | 0.5074 | 1.114 | 465.4 |
| 51.23 | 723.0 | 4.623 | −3.578 | 1.225 | 0.5072 | 1.173 | 429.9 |
| 54.93 | 757.9 | 4.656 | −3.600 | 1.206 | 0.5067 | 1.237 | 398.3 |
| 58.81 | 793.7 | 4.692 | −3.626 | 1.186 | 0.5062 | 1.307 | 370.0 |
| 62.87 | 830.4 | 4.730 | −3.658 | 1.166 | 0.5054 | 1.384 | 344.8 |
| 67.11 | 867.9 | 4.771 | −3.695 | 1.145 | 0.5044 | 1.470 | 322.1 |

Table 2, part 3 (cont.)

| t | p | $\alpha_P'$ | $\alpha_P''$ | $\delta_h'$ | $\delta_h''$ | $\delta_T'$ | $\delta_T''$ | $\frac{dr}{dT}$ |
|---|---|---|---|---|---|---|---|---|
| °C | MPa | $10^{-3}/K$ | | K/MPa | | $10^{-3}m^3/kg$ | | kJ/kg K |
| 255.0 | 4.3202 | 2.038 | 4.407 | 0.0196 | 15.59 | −0.0962 | −61.01 | −5.356 |
| 260.0 | 4.6894 | 2.127 | 4.586 | 0.0344 | 14.98 | −0.1710 | −60.96 | −5.535 |
| 265.0 | 5.0823 | 2.225 | 4.783 | 0.0504 | 14.41 | −0.2542 | −61.02 | −5.726 |
| 270.0 | 5.4999 | 2.331 | 5.002 | 0.0679 | 13.85 | −0.3469 | −61.19 | −5.929 |
| 275.0 | 5.9431 | 2.449 | 5.246 | 0.0869 | 13.32 | −0.4508 | −61.48 | −6.146 |
| 280.0 | 6.4132 | 2.578 | 5.519 | 0.1076 | 12.80 | −0.5680 | −61.92 | −6.379 |
| 285.0 | 6.9111 | 2.723 | 5.825 | 0.1304 | 12.31 | −0.7008 | −62.50 | −6.631 |
| 290.0 | 7.4380 | 2.884 | 6.170 | 0.1554 | 11.84 | −0.8525 | −63.26 | −6.904 |
| 295.0 | 7.9952 | 3.066 | 6.562 | 0.1831 | 11.38 | −1.027 | −64.21 | −7.200 |
| 300.0 | 8.5838 | 3.273 | 7.010 | 0.2139 | 10.93 | −1.229 | −65.39 | −7.525 |
| 305.0 | 9.2051 | 3.510 | 7.526 | 0.2484 | 10.50 | −1.466 | −66.83 | −7.882 |
| 310.0 | 9.8605 | 3.785 | 8.127 | 0.2871 | 10.09 | −1.747 | −68.58 | −8.279 |
| 315.0 | 10.551 | 4.107 | 8.833 | 0.3310 | 9.681 | −2.083 | −70.72 | −8.723 |
| 320.0 | 11.279 | 4.491 | 9.674 | 0.3811 | 9.285 | −2.493 | −73.33 | −9.225 |
| 325.0 | 12.046 | 4.956 | 10.69 | 0.4388 | 8.897 | −3.001 | −76.54 | −9.800 |
| 330.0 | 12.852 | 5.530 | 11.94 | 0.5060 | 8.514 | −3.644 | −80.53 | −10.47 |
| 335.0 | 13.701 | 6.258 | 13.51 | 0.5852 | 8.136 | −4.479 | −85.57 | −11.25 |
| 340.0 | 14.594 | 7.210 | 15.55 | 0.6798 | 7.759 | −5.600 | −92.06 | −12.21 |
| 345.0 | 15.533 | 8.505 | 18.28 | 0.7950 | 7.380 | −7.171 | −100.7 | −13.39 |
| 350.0 | 16.521 | 10.37 | 22.12 | 0.9382 | 6.994 | −9.500 | −112.7 | −14.92 |
| 355.0 | 17.561 | 13.25 | 27.94 | 1.121 | 6.596 | −13.24 | −130.4 | −17.03 |
| 360.0 | 18.655 | 18.30 | 37.71 | 1.364 | 6.174 | −20.04 | −159.2 | −20.17 |
| 365.0 | 19.809 | 29.10 | 57.65 | 1.706 | 5.708 | −35.36 | −215.8 | −25.72 |
| 370.0 | 21.030 | 68.20 | 126.7 | 2.255 | 5.104 | −94.60 | −401.9 | −40.47 |
| 371.0 | 21.283 | 93.92 | 170.0 | 2.429 | 4.934 | −135.0 | −515.1 | −47.96 |
| 372.0 | 21.539 | 148.8 | 274.5 | 2.654 | 4.707 | −223.1 | −784.1 | −62.21 |
| 373.0 | 21.799 | 372.2 | 683.8 | 3.011 | 4.387 | −593.1 | −1802 | −102.8 |
| 374.0 | 22.055 | ∞ | | 3.733 | | −∞ | | −∞ |

| $\dfrac{dp}{dT}\bigg|_{sat}$ | $\dfrac{d^2p}{dT^2}\bigg|_{sat}$ | $c'_{sat}$ | $c''_{sat}$ | $c'$ | $c''$ | $\chi'_T$ | $\chi''_T$ |
|---|---|---|---|---|---|---|---|
| kPa/K | Pa/K$^2$ | kJ/kg K | | km/s | | 1/GPa | |
| 71.55 | 906.4 | 4.815 | −3.739 | 1.124 | 0.5033 | 1.564 | 301.9 |
| 76.18 | 945.9 | 4.863 | −3.790 | 1.102 | 0.5019 | 1.669 | 283.8 |
| 81.01 | 986.4 | 4.914 | −3.847 | 1.080 | 0.5004 | 1.787 | 267.5 |
| 86.04 | 1028 | 4.969 | −3.914 | 1.057 | 0.4986 | 1.919 | 253.0 |
| 91.29 | 1071 | 5.030 | −3.989 | 1.033 | 0.4966 | 2.067 | 240.0 |
| 96.76 | 1115 | 5.095 | −4.074 | 1.009 | 0.4943 | 2.236 | 228.5 |
| 102.4 | 1161 | 5.167 | −4.171 | 0.9844 | 0.4918 | 2.430 | 218.3 |
| 108.4 | 1208 | 5.245 | −4.281 | 0.9589 | 0.4891 | 2.652 | 209.4 |
| 114.5 | 1258 | 5.331 | −4.406 | 0.9327 | 0.4861 | 2.909 | 201.6 |
| 121.0 | 1309 | 5.427 | −4.548 | 0.9058 | 0.4827 | 3.211 | 195.0 |
| 127.6 | 1363 | 5.534 | −4.712 | 0.8780 | 0.4791 | 3.567 | 189.5 |
| 134.6 | 1419 | 5.654 | −4.899 | 0.8494 | 0.4752 | 3.993 | 185.1 |
| 141.8 | 1479 | 5.790 | −5.116 | 0.8199 | 0.4708 | 4.509 | 181.9 |
| 149.4 | 1542 | 5.946 | −5.368 | 0.7894 | 0.4662 | 5.145 | 180.0 |
| 157.3 | 1610 | 6.127 | −5.664 | 0.7579 | 0.4610 | 5.941 | 179.5 |
| 165.5 | 1683 | 6.340 | −6.017 | 0.7253 | 0.4554 | 6.962 | 180.7 |
| 174.1 | 1762 | 6.597 | −6.443 | 0.6914 | 0.4493 | 8.307 | 183.9 |
| 183.1 | 1850 | 6.913 | −6.969 | 0.6561 | 0.4426 | 10.14 | 189.9 |
| 192.6 | 1949 | 7.315 | −7.634 | 0.6192 | 0.4351 | 12.74 | 199.6 |
| 202.6 | 2064 | 7.848 | −8.509 | 0.5806 | 0.4267 | 16.67 | 215.2 |
| 213.3 | 2201 | 8.599 | −9.722 | 0.5398 | 0.4172 | 23.09 | 240.7 |
| 224.7 | 2375 | 9.760 | −11.55 | 0.4966 | 0.4061 | 35.00 | 285.2 |
| 237.2 | 2631 | 11.87 | −14.80 | 0.4502 | 0.3928 | 62.40 | 377.3 |
| 251.4 | 3132 | 17.62 | −23.55 | 0.3961 | 0.3743 | 171.4 | 690.5 |
| 254.6 | 3297 | 20.63 | −27.96 | 0.3814 | 0.3691 | 247.5 | 881.4 |
| 258.5 | 3729 | 25.88 | −36.87 | 0.3617 | 0.3605 | 417.0 | 1333 |
| 262.6 | 4801 | 40.98 | −62.21 | 0.3255 | 0.3410 | 1149 | 3042 |
| 267.9 | − | ∞ | −∞ | − | | ∞ | |

Table 2, part 4

| $p$ | $t$ | $\varrho'$ | $\varrho''$ | $h'$ | $h''$ | $r$ |
|---|---|---|---|---|---|---|
| MPa | °C | kg/m³ | | kJ/kg | | kJ/kg |
| 0.00061 | 0.01 | 999.8 | 0.0049 | 0.0006 | 2500.5 | 2500.5 |
| 0.00065 | 0.8 | 999.8 | 0.0051 | 3.5421 | 2502.1 | 2498.5 |
| 0.00070 | 1.9 | 999.9 | 0.0055 | 7.8939 | 2504.0 | 2496.1 |
| 0.00075 | 2.8 | 999.9 | 0.0059 | 11.973 | 2505.8 | 2493.8 |
| 0.00080 | 3.8 | 999.9 | 0.0063 | 15.813 | 2507.4 | 2491.6 |
| 0.00085 | 4.6 | 999.9 | 0.0066 | 19.443 | 2509.0 | 2489.6 |
| 0.00090 | 5.4 | 999.9 | 0.0070 | 22.886 | 2510.5 | 2487.6 |
| 0.00095 | 6.2 | 999.9 | 0.0074 | 26.162 | 2512.0 | 2485.8 |
| 0.00100 | 7.0 | 999.9 | 0.0077 | 29.288 | 2513.3 | 2484.0 |
| 0.00110 | 8.4 | 999.8 | 0.0085 | 35.142 | 2515.9 | 2480.7 |
| 0.00120 | 9.7 | 999.7 | 0.0092 | 40.541 | 2518.3 | 2477.7 |
| 0.00130 | 10.9 | 999.6 | 0.0099 | 45.555 | 2520.4 | 2474.9 |
| 0.00140 | 12.0 | 999.5 | 0.0106 | 50.238 | 2522.5 | 2472.3 |
| 0.00150 | 13.0 | 999.4 | 0.0114 | 54.635 | 2524.4 | 2469.8 |
| 0.00160 | 14.0 | 999.2 | 0.0121 | 58.781 | 2526.2 | 2467.5 |
| 0.00170 | 14.9 | 999.1 | 0.0128 | 62.705 | 2527.9 | 2465.2 |
| 0.00180 | 15.8 | 999.0 | 0.0135 | 66.430 | 2529.6 | 2463.1 |
| 0.00190 | 16.7 | 998.8 | 0.0142 | 69.978 | 2531.1 | 2461.1 |
| 0.00200 | 17.5 | 998.7 | 0.0149 | 73.366 | 2532.6 | 2459.2 |
| 0.00220 | 19.0 | 998.4 | 0.0163 | 79.717 | 2535.4 | 2455.7 |
| 0.00240 | 20.4 | 998.1 | 0.0177 | 85.582 | 2537.9 | 2452.3 |
| 0.00260 | 21.7 | 997.8 | 0.0191 | 91.034 | 2540.3 | 2449.3 |
| 0.00280 | 22.9 | 997.5 | 0.0205 | 96.130 | 2542.5 | 2446.4 |
| 0.00300 | 24.1 | 997.3 | 0.0219 | 100.92 | 2544.6 | 2443.7 |
| 0.00320 | 25.2 | 997.0 | 0.0233 | 105.44 | 2546.6 | 2441.1 |
| 0.00340 | 26.2 | 996.7 | 0.0246 | 109.71 | 2548.4 | 2438.7 |
| 0.00360 | 27.2 | 996.4 | 0.0260 | 113.78 | 2550.2 | 2436.4 |
| 0.00380 | 28.1 | 996.2 | 0.0274 | 117.65 | 2551.9 | 2434.2 |
| 0.00400 | 29.0 | 995.9 | 0.0287 | 121.35 | 2553.5 | 2432.1 |
| 0.00450 | 31.0 | 995.3 | 0.0321 | 129.93 | 2557.2 | 2427.3 |

| $u'$ | $u''$ | $s'$ | $s''$ | $c_p'$ | $c_p''$ | $c_v'$ | $c_v''$ |
|------|-------|------|-------|--------|---------|--------|---------|
| kJ/kg | | kJ/kg K | | kJ/kg K | | kJ/kg K | |
| 0.0000 | 2374.5 | 0.000 | 9.154 | 4.229 | 1.868 | 4.225 | 1.404 |
| 3.5415 | 2375.7 | 0.013 | 9.132 | 4.222 | 1.868 | 4.220 | 1.404 |
| 7.8932 | 2377.1 | 0.029 | 9.104 | 4.215 | 1.869 | 4.214 | 1.405 |
| 11.972 | 2378.4 | 0.044 | 9.079 | 4.210 | 1.869 | 4.209 | 1.405 |
| 15.812 | 2379.7 | 0.057 | 9.055 | 4.205 | 1.870 | 4.205 | 1.405 |
| 19.442 | 2380.9 | 0.071 | 9.033 | 4.202 | 1.870 | 4.202 | 1.406 |
| 22.885 | 2382.0 | 0.083 | 9.012 | 4.199 | 1.871 | 4.198 | 1.406 |
| 26.161 | 2383.1 | 0.095 | 8.992 | 4.196 | 1.871 | 4.196 | 1.407 |
| 29.287 | 2384.1 | 0.106 | 8.974 | 4.194 | 1.872 | 4.193 | 1.407 |
| 35.141 | 2386.1 | 0.127 | 8.939 | 4.191 | 1.873 | 4.189 | 1.408 |
| 40.540 | 2387.8 | 0.146 | 8.907 | 4.189 | 1.874 | 4.185 | 1.409 |
| 45.553 | 2389.5 | 0.164 | 8.878 | 4.187 | 1.875 | 4.181 | 1.409 |
| 50.237 | 2391.0 | 0.180 | 8.851 | 4.186 | 1.875 | 4.178 | 1.410 |
| 54.634 | 2392.5 | 0.195 | 8.826 | 4.185 | 1.876 | 4.175 | 1.411 |
| 58.779 | 2393.8 | 0.210 | 8.802 | 4.185 | 1.877 | 4.173 | 1.411 |
| 62.703 | 2395.1 | 0.223 | 8.780 | 4.184 | 1.878 | 4.170 | 1.412 |
| 66.428 | 2396.3 | 0.236 | 8.760 | 4.184 | 1.878 | 4.168 | 1.412 |
| 69.976 | 2397.5 | 0.249 | 8.740 | 4.184 | 1.879 | 4.165 | 1.413 |
| 73.364 | 2398.6 | 0.260 | 8.722 | 4.184 | 1.880 | 4.163 | 1.413 |
| 79.715 | 2400.7 | 0.282 | 8.687 | 4.183 | 1.881 | 4.158 | 1.414 |
| 85.580 | 2402.6 | 0.302 | 8.656 | 4.183 | 1.882 | 4.154 | 1.415 |
| 91.031 | 2404.4 | 0.321 | 8.627 | 4.183 | 1.884 | 4.150 | 1.416 |
| 96.127 | 2406.1 | 0.338 | 8.600 | 4.183 | 1.885 | 4.146 | 1.417 |
| 100.92 | 2407.6 | 0.354 | 8.576 | 4.183 | 1.886 | 4.142 | 1.418 |
| 105.43 | 2409.1 | 0.369 | 8.552 | 4.183 | 1.887 | 4.138 | 1.419 |
| 109.71 | 2410.5 | 0.384 | 8.531 | 4.183 | 1.888 | 4.135 | 1.420 |
| 113.77 | 2411.8 | 0.397 | 8.510 | 4.183 | 1.889 | 4.131 | 1.420 |
| 117.64 | 2413.1 | 0.410 | 8.491 | 4.183 | 1.890 | 4.128 | 1.421 |
| 121.34 | 2414.3 | 0.422 | 8.473 | 4.183 | 1.891 | 4.124 | 1.422 |
| 129.93 | 2417.1 | 0.451 | 8.431 | 4.183 | 1.893 | 4.116 | 1.424 |

| $p$ | $t$ | $\varrho\,'$ | $\varrho\,''$ | $h\,'$ | $h\,''$ | $r$ |
|---|---|---|---|---|---|---|
| MPa | °C | kg/m³ | | kJ/kg | | kJ/kg |
| 0.00500 | 32.88 | 994.7 | 0.0355 | 137.72 | 2560.5 | 2422.8 |
| 0.00550 | 34.59 | 994.1 | 0.0388 | 144.87 | 2563.6 | 2418.8 |
| 0.00600 | 36.17 | 993.6 | 0.0421 | 151.47 | 2566.5 | 2415.0 |
| 0.00650 | 37.63 | 993.1 | 0.0454 | 157.61 | 2569.1 | 2411.5 |
| 0.00700 | 39.01 | 992.5 | 0.0487 | 163.35 | 2571.6 | 2408.2 |
| 0.00750 | 40.30 | 992.1 | 0.0520 | 168.76 | 2573.9 | 2405.1 |
| 0.00800 | 41.52 | 991.6 | 0.0552 | 173.85 | 2576.1 | 2402.2 |
| 0.00850 | 42.67 | 991.1 | 0.0585 | 178.68 | 2578.1 | 2399.5 |
| 0.00900 | 43.77 | 990.7 | 0.0617 | 183.27 | 2580.1 | 2396.8 |
| 0.00950 | 44.82 | 990.2 | 0.0649 | 187.65 | 2582.0 | 2394.3 |
| 0.01000 | 45.82 | 989.8 | 0.0681 | 191.83 | 2583.8 | 2391.9 |
| 0.01200 | 49.43 | 988.2 | 0.0809 | 206.95 | 2590.2 | 2383.2 |
| 0.01400 | 52.56 | 986.8 | 0.0935 | 220.03 | 2595.7 | 2375.7 |
| 0.01600 | 55.33 | 985.5 | 0.1060 | 231.61 | 2600.6 | 2369.0 |
| 0.01800 | 57.81 | 984.3 | 0.1184 | 242.00 | 2605.0 | 2363.0 |
| 0.02000 | 60.07 | 983.1 | 0.1307 | 251.46 | 2608.9 | 2357.5 |
| 0.02200 | 62.15 | 982.0 | 0.1430 | 260.15 | 2612.5 | 2352.4 |
| 0.02400 | 64.07 | 981.0 | 0.1551 | 268.18 | 2615.9 | 2347.7 |
| 0.02600 | 65.86 | 980.1 | 0.1672 | 275.67 | 2619.0 | 2343.3 |
| 0.02800 | 67.54 | 979.1 | 0.1792 | 282.69 | 2621.9 | 2339.2 |
| 0.03000 | 69.11 | 978.3 | 0.1912 | 289.30 | 2624.6 | 2335.3 |
| 0.03500 | 72.70 | 976.2 | 0.2209 | 304.32 | 2630.7 | 2326.4 |
| 0.04000 | 75.88 | 974.3 | 0.2504 | 317.64 | 2636.1 | 2318.5 |
| 0.04500 | 78.74 | 972.6 | 0.2796 | 329.62 | 2640.9 | 2311.3 |
| 0.05000 | 81.34 | 971.0 | 0.3086 | 340.54 | 2645.3 | 2304.8 |
| 0.05500 | 83.73 | 969.4 | 0.3374 | 350.59 | 2649.3 | 2298.7 |
| 0.06000 | 85.95 | 968.0 | 0.3660 | 359.90 | 2653.0 | 2293.1 |
| 0.06500 | 88.02 | 966.7 | 0.3944 | 368.60 | 2656.4 | 2287.8 |
| 0.07000 | 89.96 | 965.4 | 0.4228 | 376.75 | 2659.6 | 2282.8 |
| 0.07500 | 91.78 | 964.1 | 0.4510 | 384.43 | 2662.5 | 2278.1 |

| $u'$ | $u''$ | $s'$ | $s''$ | $c_p'$ | $c_p''$ | $c_v'$ | $c_v''$ |
|---|---|---|---|---|---|---|---|
| kJ/kg | | kJ/kg K | | kJ/kg K | | kJ/kg K | |
| 137.72 | 2419.6 | 0.476 | 8.393 | 4.183 | 1.895 | 4.108 | 1.425 |
| 144.86 | 2421.9 | 0.499 | 8.359 | 4.183 | 1.897 | 4.101 | 1.427 |
| 151.46 | 2424.0 | 0.521 | 8.328 | 4.183 | 1.899 | 4.094 | 1.428 |
| 157.60 | 2426.0 | 0.541 | 8.300 | 4.183 | 1.901 | 4.087 | 1.430 |
| 163.35 | 2427.9 | 0.559 | 8.274 | 4.183 | 1.903 | 4.081 | 1.431 |
| 168.75 | 2429.6 | 0.576 | 8.249 | 4.182 | 1.905 | 4.075 | 1.432 |
| 173.85 | 2431.3 | 0.593 | 8.227 | 4.182 | 1.907 | 4.069 | 1.434 |
| 178.68 | 2432.8 | 0.608 | 8.205 | 4.182 | 1.908 | 4.064 | 1.435 |
| 183.27 | 2434.3 | 0.622 | 8.185 | 4.182 | 1.910 | 4.058 | 1.436 |
| 187.64 | 2435.7 | 0.636 | 8.166 | 4.182 | 1.911 | 4.053 | 1.437 |
| 191.82 | 2437.0 | 0.649 | 8.148 | 4.182 | 1.913 | 4.048 | 1.438 |
| 206.93 | 2441.8 | 0.696 | 8.084 | 4.182 | 1.918 | 4.030 | 1.442 |
| 220.02 | 2446.0 | 0.737 | 8.031 | 4.182 | 1.924 | 4.014 | 1.446 |
| 231.59 | 2449.7 | 0.772 | 7.984 | 4.182 | 1.929 | 4.000 | 1.449 |
| 241.99 | 2452.9 | 0.804 | 7.943 | 4.182 | 1.933 | 3.987 | 1.453 |
| 251.44 | 2455.9 | 0.832 | 7.907 | 4.183 | 1.937 | 3.975 | 1.456 |
| 260.12 | 2458.6 | 0.858 | 7.874 | 4.183 | 1.941 | 3.964 | 1.458 |
| 268.16 | 2461.1 | 0.882 | 7.844 | 4.184 | 1.945 | 3.954 | 1.461 |
| 275.65 | 2463.5 | 0.904 | 7.816 | 4.185 | 1.949 | 3.945 | 1.464 |
| 282.66 | 2465.6 | 0.925 | 7.791 | 4.186 | 1.953 | 3.936 | 1.466 |
| 289.27 | 2467.7 | 0.944 | 7.767 | 4.186 | 1.956 | 3.928 | 1.469 |
| 304.28 | 2472.3 | 0.988 | 7.714 | 4.189 | 1.964 | 3.910 | 1.474 |
| 317.59 | 2476.4 | 1.026 | 7.669 | 4.191 | 1.972 | 3.893 | 1.479 |
| 329.58 | 2480.0 | 1.060 | 7.629 | 4.193 | 1.979 | 3.878 | 1.484 |
| 340.49 | 2483.3 | 1.091 | 7.593 | 4.195 | 1.986 | 3.865 | 1.489 |
| 350.53 | 2486.3 | 1.119 | 7.561 | 4.198 | 1.993 | 3.853 | 1.493 |
| 359.84 | 2489.0 | 1.145 | 7.531 | 4.200 | 1.999 | 3.842 | 1.497 |
| 368.53 | 2491.6 | 1.170 | 7.504 | 4.202 | 2.005 | 3.831 | 1.501 |
| 376.68 | 2494.0 | 1.192 | 7.479 | 4.204 | 2.011 | 3.821 | 1.505 |
| 384.36 | 2496.2 | 1.213 | 7.456 | 4.206 | 2.017 | 3.812 | 1.509 |

| $p$ | $t$ | $\varrho'$ | $\varrho''$ | $h'$ | $h''$ | $r$ |
|------|------|------|------|------|------|------|
| MPa | °C | kg/m³ | | kJ/kg | | kJ/kg |
| 0.08000 | 93.51 | 962.9 | 0.4790 | 391.71 | 2665.3 | 2273.6 |
| 0.09000 | 96.71 | 960.7 | 0.5348 | 405.20 | 2670.5 | 2265.3 |
| 0.10000 | 99.63 | 958.7 | 0.5902 | 417.51 | 2675.1 | 2257.6 |
| 0.1200 | 104.81 | 954.9 | 0.6999 | 439.38 | 2683.3 | 2243.9 |
| 0.1400 | 109.32 | 951.5 | 0.8085 | 458.46 | 2690.2 | 2231.8 |
| 0.1600 | 113.33 | 948.4 | 0.9161 | 475.44 | 2696.3 | 2220.9 |
| 0.1800 | 116.94 | 945.6 | 1.0228 | 490.78 | 2701.7 | 2210.9 |
| 0.2000 | 120.24 | 943.0 | 1.1289 | 504.80 | 2706.5 | 2201.7 |
| 0.2500 | 127.44 | 937.0 | 1.3912 | 535.49 | 2716.8 | 2181.4 |
| 0.3000 | 133.56 | 931.8 | 1.6505 | 561.61 | 2725.3 | 2163.7 |
| 0.3500 | 138.89 | 927.2 | 1.9074 | 584.48 | 2732.4 | 2147.9 |
| 0.4000 | 143.64 | 922.9 | 2.1624 | 604.91 | 2738.5 | 2133.6 |
| 0.4500 | 147.94 | 919.0 | 2.4157 | 623.42 | 2743.9 | 2120.5 |
| 0.5000 | 151.87 | 915.3 | 2.6677 | 640.38 | 2748.6 | 2108.2 |
| 0.6000 | 158.86 | 908.6 | 3.1683 | 670.71 | 2756.7 | 2086.0 |
| 0.7000 | 164.98 | 902.6 | 3.6655 | 697.35 | 2763.3 | 2066.0 |
| 0.8000 | 170.44 | 897.1 | 4.1603 | 721.23 | 2768.9 | 2047.7 |
| 0.9000 | 175.39 | 891.9 | 4.6531 | 742.93 | 2773.6 | 2030.7 |
| 1.0000 | 179.92 | 887.2 | 5.1445 | 762.88 | 2777.7 | 2014.8 |
| 1.1000 | 184.10 | 882.6 | 5.6349 | 781.38 | 2781.2 | 1999.9 |
| 1.2000 | 188.00 | 878.4 | 6.1246 | 798.68 | 2784.3 | 1985.7 |
| 1.3000 | 191.64 | 874.3 | 6.6138 | 814.93 | 2787.0 | 1972.1 |
| 1.4000 | 195.08 | 870.4 | 7.1028 | 830.28 | 2789.4 | 1959.1 |
| 1.5000 | 198.33 | 866.7 | 7.5918 | 844.85 | 2791.5 | 1946.7 |
| 1.6000 | 201.41 | 863.1 | 8.0809 | 858.73 | 2793.3 | 1934.6 |
| 1.7000 | 204.35 | 859.6 | 8.5703 | 872.00 | 2795.0 | 1923.0 |
| 1.8000 | 207.15 | 856.3 | 9.0601 | 884.71 | 2796.4 | 1911.7 |
| 1.9000 | 209.84 | 853.0 | 9.5504 | 896.92 | 2797.6 | 1900.7 |
| 2.0000 | 212.42 | 849.9 | 10.041 | 908.69 | 2798.7 | 1890.0 |
| 2.2000 | 217.29 | 843.8 | 11.025 | 931.01 | 2800.5 | 1869.5 |

| $u'$ | $u''$ | $s'$ | $s''$ | $c_p'$ | $c_p''$ | $c_v'$ | $c_v''$ |
|---|---|---|---|---|---|---|---|
| kJ/kg | | kJ/kg K | | kJ/kg K | | kJ/kg K | |
| 391.63 | 2498.3 | 1.233 | 7.434 | 4.209 | 2.022 | 3.803 | 1.512 |
| 405.11 | 2502.2 | 1.270 | 7.394 | 4.213 | 2.033 | 3.787 | 1.519 |
| 417.41 | 2505.7 | 1.303 | 7.359 | 4.217 | 2.043 | 3.773 | 1.525 |
| 439.26 | 2511.8 | 1.361 | 7.298 | 4.224 | 2.062 | 3.747 | 1.537 |
| 458.31 | 2517.1 | 1.411 | 7.246 | 4.231 | 2.079 | 3.725 | 1.548 |
| 475.27 | 2521.7 | 1.455 | 7.202 | 4.237 | 2.096 | 3.705 | 1.558 |
| 490.59 | 2525.7 | 1.495 | 7.162 | 4.243 | 2.112 | 3.687 | 1.568 |
| 504.59 | 2529.4 | 1.530 | 7.127 | 4.249 | 2.127 | 3.671 | 1.577 |
| 535.22 | 2537.1 | 1.608 | 7.053 | 4.262 | 2.162 | 3.636 | 1.597 |
| 561.29 | 2543.5 | 1.672 | 6.992 | 4.275 | 2.195 | 3.607 | 1.616 |
| 584.10 | 2548.9 | 1.728 | 6.941 | 4.286 | 2.226 | 3.582 | 1.634 |
| 604.47 | 2553.5 | 1.777 | 6.896 | 4.297 | 2.256 | 3.559 | 1.651 |
| 622.93 | 2557.6 | 1.821 | 6.857 | 4.307 | 2.284 | 3.539 | 1.666 |
| 639.84 | 2561.2 | 1.861 | 6.821 | 4.317 | 2.312 | 3.521 | 1.682 |
| 670.05 | 2567.3 | 1.932 | 6.760 | 4.335 | 2.365 | 3.489 | 1.710 |
| 696.58 | 2572.4 | 1.993 | 6.708 | 4.353 | 2.415 | 3.462 | 1.736 |
| 720.33 | 2576.6 | 2.046 | 6.663 | 4.370 | 2.464 | 3.438 | 1.761 |
| 741.92 | 2580.2 | 2.095 | 6.622 | 4.387 | 2.511 | 3.417 | 1.785 |
| 761.75 | 2583.3 | 2.139 | 6.586 | 4.403 | 2.557 | 3.398 | 1.808 |
| 780.14 | 2586.0 | 2.179 | 6.553 | 4.419 | 2.602 | 3.380 | 1.830 |
| 797.31 | 2588.4 | 2.217 | 6.523 | 4.435 | 2.646 | 3.364 | 1.852 |
| 813.44 | 2590.5 | 2.252 | 6.495 | 4.450 | 2.689 | 3.350 | 1.873 |
| 828.67 | 2592.3 | 2.284 | 6.468 | 4.466 | 2.732 | 3.336 | 1.893 |
| 843.12 | 2593.9 | 2.315 | 6.444 | 4.481 | 2.775 | 3.324 | 1.912 |
| 856.88 | 2595.3 | 2.344 | 6.421 | 4.496 | 2.816 | 3.312 | 1.931 |
| 870.02 | 2596.6 | 2.372 | 6.399 | 4.511 | 2.858 | 3.301 | 1.950 |
| 882.61 | 2597.7 | 2.398 | 6.378 | 4.526 | 2.899 | 3.290 | 1.969 |
| 894.70 | 2598.7 | 2.423 | 6.358 | 4.541 | 2.940 | 3.281 | 1.987 |
| 906.33 | 2599.5 | 2.447 | 6.340 | 4.556 | 2.981 | 3.271 | 2.004 |
| 928.41 | 2600.9 | 2.492 | 6.304 | 4.586 | 3.062 | 3.254 | 2.039 |

| $p$ MPa | $t$ °C | $\varrho'$ | $\varrho''$ | $h'$ | $h''$ | $r$ |
|---|---|---|---|---|---|---|
| | | kg/m³ | | kJ/kg | | kJ/kg |
| 2.4000 | 221.83 | 838.0 | 12.013 | 951.96 | 2801.7 | 1849.8 |
| 2.6000 | 226.08 | 832.5 | 13.004 | 971.71 | 2802.6 | 1830.9 |
| 2.8000 | 230.10 | 827.1 | 14.000 | 990.45 | 2803.1 | 1812.6 |
| 3.0000 | 233.89 | 822.0 | 15.001 | 1008.3 | 2803.3 | 1795.0 |
| 3.2000 | 237.50 | 817.0 | 16.007 | 1025.3 | 2803.2 | 1777.8 |
| 3.4000 | 240.93 | 812.2 | 17.019 | 1041.7 | 2802.8 | 1761.1 |
| 3.6000 | 244.22 | 807.5 | 18.037 | 1057.4 | 2802.3 | 1744.9 |
| 3.8000 | 247.37 | 802.9 | 19.061 | 1072.6 | 2801.5 | 1729.0 |
| 4.0000 | 250.39 | 798.5 | 20.092 | 1087.2 | 2800.6 | 1713.4 |
| 4.5000 | 257.47 | 787.8 | 22.700 | 1121.9 | 2797.6 | 1675.7 |
| 5.0000 | 263.98 | 777.5 | 25.355 | 1154.2 | 2793.7 | 1639.5 |
| 6.0000 | 275.62 | 758.2 | 30.825 | 1213.3 | 2783.9 | 1570.6 |
| 7.0000 | 285.86 | 739.9 | 36.534 | 1267.0 | 2771.8 | 1504.8 |
| 8.0000 | 295.04 | 722.4 | 42.518 | 1316.6 | 2757.8 | 1441.2 |
| 9.0000 | 303.38 | 705.4 | 48.817 | 1363.1 | 2742.0 | 1378.9 |
| 10.000 | 311.03 | 688.6 | 55.477 | 1407.3 | 2724.5 | 1317.2 |
| 11.000 | 318.11 | 672.0 | 62.555 | 1449.7 | 2705.4 | 1255.7 |
| 12.000 | 324.71 | 655.3 | 70.118 | 1490.7 | 2684.5 | 1193.8 |
| 13.000 | 330.89 | 638.5 | 78.253 | 1530.9 | 2661.8 | 1131.0 |
| 14.000 | 336.70 | 621.3 | 87.071 | 1570.4 | 2637.1 | 1066.7 |
| 15.000 | 342.19 | 603.5 | 96.720 | 1609.8 | 2610.1 | 1000.2 |
| 17.500 | 354.72 | 554.5 | 126.08 | 1710.7 | 2529.0 | 818.29 |
| 20.000 | 365.80 | 491.2 | 170.25 | 1826.7 | 2413.6 | 586.81 |
| 21.000 | 369.88 | 454.5 | 199.19 | 1887.6 | 2342.8 | 455.19 |
| 21.200 | 370.67 | 445.0 | 207.03 | 1902.6 | 2324.3 | 421.67 |
| 21.400 | 371.46 | 434.1 | 216.20 | 1919.5 | 2302.9 | 383.40 |
| 21.600 | 372.24 | 420.9 | 227.46 | 1939.6 | 2277.1 | 337.47 |
| 21.800 | 373.00 | 402.3 | 242.76 | 1966.8 | 2242.8 | 276.00 |
| 22.000 | 373.8 | 370.2 | 273.78 | 2012.7 | 2176.3 | 163.62 |
| 22.055 | 374.0 | 322.0 | | 2085.8 | | 0 |

| $u'$ | $u''$ | $s'$ | $s''$ | $c_p'$ | $c_p''$ | $c_v'$ | $c_v''$ |
|---|---|---|---|---|---|---|---|
| kJ/kg | | kJ/kg K | | kJ/kg K | | kJ/kg K | |
| 949.09 | 2601.9 | 2.534 | 6.272 | 4.616 | 3.142 | 3.239 | 2.072 |
| 968.59 | 2602.6 | 2.574 | 6.241 | 4.646 | 3.222 | 3.225 | 2.104 |
| 987.06 | 2603.1 | 2.611 | 6.212 | 4.676 | 3.301 | 3.212 | 2.136 |
| 1004.6 | 2603.3 | 2.645 | 6.186 | 4.706 | 3.381 | 3.200 | 2.166 |
| 1021.4 | 2603.3 | 2.678 | 6.160 | 4.737 | 3.461 | 3.189 | 2.196 |
| 1037.5 | 2603.1 | 2.710 | 6.136 | 4.767 | 3.541 | 3.179 | 2.226 |
| 1053.0 | 2602.7 | 2.740 | 6.112 | 4.798 | 3.621 | 3.170 | 2.254 |
| 1067.8 | 2602.2 | 2.769 | 6.090 | 4.829 | 3.702 | 3.161 | 2.282 |
| 1082.2 | 2601.5 | 2.796 | 6.069 | 4.861 | 3.783 | 3.153 | 2.310 |
| 1116.2 | 2599.4 | 2.861 | 6.019 | 4.942 | 3.989 | 3.135 | 2.377 |
| 1147.8 | 2596.5 | 2.920 | 5.973 | 5.025 | 4.200 | 3.120 | 2.441 |
| 1205.4 | 2589.3 | 3.027 | 5.889 | 5.201 | 4.643 | 3.096 | 2.563 |
| 1257.5 | 2580.2 | 3.121 | 5.813 | 5.394 | 5.121 | 3.078 | 2.679 |
| 1305.5 | 2569.6 | 3.207 | 5.743 | 5.609 | 5.647 | 3.067 | 2.789 |
| 1350.3 | 2557.6 | 3.285 | 5.677 | 5.849 | 6.233 | 3.059 | 2.895 |
| 1392.8 | 2544.3 | 3.359 | 5.614 | 6.124 | 6.897 | 3.056 | 2.998 |
| 1433.3 | 2529.5 | 3.429 | 5.553 | 6.443 | 7.662 | 3.057 | 3.097 |
| 1472.4 | 2513.4 | 3.495 | 5.492 | 6.820 | 8.558 | 3.061 | 3.195 |
| 1510.5 | 2495.7 | 3.559 | 5.432 | 7.274 | 9.629 | 3.069 | 3.291 |
| 1547.9 | 2476.3 | 3.622 | 5.371 | 7.836 | 10.94 | 3.081 | 3.386 |
| 1585.0 | 2455.0 | 3.684 | 5.309 | 8.551 | 12.58 | 3.099 | 3.480 |
| 1679.2 | 2390.2 | 3.839 | 5.142 | 11.69 | 19.51 | 3.169 | 3.713 |
| 1786.0 | 2296.1 | 4.015 | 4.933 | 22.37 | 40.99 | 3.313 | 3.926 |
| 1841.4 | 2237.4 | 4.106 | 4.814 | 40.82 | 76.58 | 3.438 | 4.059 |
| 1854.9 | 2221.9 | 4.129 | 4.784 | 50.13 | 94.20 | 3.480 | 4.104 |
| 1870.2 | 2203.9 | 4.154 | 4.749 | 66.05 | 123.7 | 3.545 | 4.175 |
| 1888.3 | 2182.1 | 4.185 | 4.708 | 99.77 | 182.4 | 3.679 | 4.327 |
| 1912.6 | 2153.0 | 4.226 | 4.653 | 221.3 | 351.1 | 4.051 | 4.685 |
| 1953.3 | 2096.0 | 4.296 | 4.549 | 1404 | 4042 | 5.273 | 5.837 |
| 2017.3 | | 4.409 | | ∞ | | — | |

Table 2, part 5

| $p$ MPa | $t$ °C | $\lambda'$ | $\lambda''$ mW/Km | $\eta'$ | $\eta''$ μPa s | $\nu'$ | $\nu''$ $10^{-6}$ m²/s |
|---|---|---|---|---|---|---|---|
| 0.00061 | 0.01 | 561.0 | 17.07 | 1792 | 9.216 | 1.792 | 1898 |
| 0.00065 | 0.8  | 562.6 | 17.11 | 1741 | 9.235 | 1.741 | 1796 |
| 0.00070 | 1.9  | 564.6 | 17.17 | 1681 | 9.260 | 1.681 | 1678 |
| 0.00075 | 2.8  | 566.4 | 17.22 | 1628 | 9.283 | 1.628 | 1576 |
| 0.00080 | 3.8  | 568.2 | 17.27 | 1580 | 9.305 | 1.580 | 1486 |
| 0.00085 | 4.6  | 569.8 | 17.32 | 1537 | 9.326 | 1.537 | 1406 |
| 0.00090 | 5.4  | 571.4 | 17.36 | 1498 | 9.346 | 1.498 | 1334 |
| 0.00095 | 6.2  | 572.8 | 17.41 | 1462 | 9.365 | 1.462 | 1270 |
| 0.00100 | 7.0  | 574.3 | 17.45 | 1429 | 9.384 | 1.429 | 1212 |
| 0.00110 | 8.4  | 576.9 | 17.53 | 1370 | 9.419 | 1.371 | 1112 |
| 0.00120 | 9.7  | 579.3 | 17.60 | 1320 | 9.452 | 1.320 | 1027 |
| 0.00130 | 10.9 | 581.6 | 17.67 | 1275 | 9.483 | 1.276 | 955.3 |
| 0.00140 | 12.0 | 583.7 | 17.74 | 1236 | 9.512 | 1.236 | 893.3 |
| 0.00150 | 13.0 | 585.7 | 17.80 | 1200 | 9.539 | 1.201 | 839.2 |
| 0.00160 | 14.0 | 587.5 | 17.86 | 1168 | 9.565 | 1.169 | 791.6 |
| 0.00170 | 14.9 | 589.2 | 17.91 | 1140 | 9.590 | 1.141 | 749.3 |
| 0.00180 | 15.8 | 590.9 | 17.97 | 1113 | 9.614 | 1.114 | 711.6 |
| 0.00190 | 16.7 | 592.4 | 18.02 | 1089 | 9.637 | 1.090 | 677.7 |
| 0.00200 | 17.5 | 593.9 | 18.07 | 1067 | 9.658 | 1.068 | 647.1 |
| 0.00220 | 19.0 | 596.6 | 18.16 | 1027 | 9.700 | 1.028 | 593.8 |
| 0.00240 | 20.4 | 599.1 | 18.25 | 991.9 | 9.738 | 0.994 | 549.1 |
| 0.00260 | 21.7 | 601.4 | 18.34 | 961.2 | 9.774 | 0.963 | 510.9 |
| 0.00280 | 22.9 | 603.6 | 18.41 | 933.9 | 9.808 | 0.936 | 478.0 |
| 0.00300 | 24.1 | 605.6 | 18.49 | 909.4 | 9.840 | 0.912 | 449.3 |
| 0.00320 | 25.2 | 607.4 | 18.56 | 887.2 | 9.871 | 0.890 | 424.1 |
| 0.00340 | 26.2 | 609.1 | 18.63 | 866.9 | 9.900 | 0.870 | 401.6 |
| 0.00360 | 27.2 | 610.8 | 18.69 | 848.4 | 9.928 | 0.851 | 381.6 |
| 0.00380 | 28.1 | 612.3 | 18.75 | 831.3 | 9.954 | 0.835 | 363.6 |
| 0.00400 | 29.0 | 613.7 | 18.81 | 815.5 | 9.980 | 0.819 | 347.3 |
| 0.00450 | 31.0 | 617.1 | 18.96 | 780.7 | 10.04 | 0.784 | 312.6 |

| $a'$ | $a''$ | $Pr'$ | $Pr''$ | $\sigma$ | $b$ | $\beta_v'$ | $\beta_s'$ |
|------|-------|-------|--------|----------|-----|-----------|-----------|
| $10^{-6}$ m²/s | | — | | mN/m | mm | 1/K | |
| 0.1327 | 1883 | 13.51 | 1.008 | 75.65 | 2.778 | −257.8 | −314540 |
| 0.1333 | 1781 | 13.06 | 1.008 | 75.53 | 2.776 | −193.6 | −371850 |
| 0.1339 | 1665 | 12.55 | 1.008 | 75.39 | 2.773 | −125.0 | −497660 |
| 0.1346 | 1564 | 12.10 | 1.008 | 75.25 | 2.770 | −69.97 | −776400 |
| 0.1351 | 1475 | 11.70 | 1.007 | 75.12 | 2.768 | −25.27 | −1893400 |
| 0.1356 | 1396 | 11.33 | 1.007 | 75.00 | 2.766 | 11.35 | 3740100 |
| 0.1361 | 1325 | 11.01 | 1.007 | 74.88 | 2.763 | 41.59 | 912160 |
| 0.1365 | 1262 | 10.71 | 1.007 | 74.77 | 2.761 | 66.72 | 511100 |
| 0.1369 | 1204 | 10.44 | 1.007 | 74.66 | 2.759 | 87.71 | 351310 |
| 0.1377 | 1105 | 9.954 | 1.006 | 74.46 | 2.756 | 120.2 | 212370 |
| 0.1384 | 1021 | 9.540 | 1.006 | 74.27 | 2.752 | 143.4 | 149860 |
| 0.1390 | 949.6 | 9.179 | 1.006 | 74.10 | 2.749 | 160.0 | 114560 |
| 0.1395 | 888.1 | 8.861 | 1.006 | 73.94 | 2.747 | 172.0 | 91982 |
| 0.1400 | 834.5 | 8.577 | 1.006 | 73.78 | 2.744 | 180.6 | 76374 |
| 0.1405 | 787.3 | 8.322 | 1.005 | 73.64 | 2.741 | 186.7 | 64978 |
| 0.1410 | 745.5 | 8.092 | 1.005 | 73.50 | 2.739 | 190.9 | 56320 |
| 0.1414 | 708.1 | 7.882 | 1.005 | 73.36 | 2.737 | 193.6 | 49535 |
| 0.1418 | 674.4 | 7.690 | 1.005 | 73.24 | 2.734 | 195.3 | 44088 |
| 0.1422 | 644.0 | 7.513 | 1.005 | 73.12 | 2.732 | 196.1 | 39627 |
| 0.1429 | 591.2 | 7.198 | 1.005 | 72.89 | 2.728 | 195.9 | 32776 |
| 0.1435 | 546.7 | 6.925 | 1.004 | 72.68 | 2.725 | 194.1 | 27784 |
| 0.1441 | 508.9 | 6.686 | 1.004 | 72.48 | 2.722 | 191.3 | 24001 |
| 0.1446 | 476.2 | 6.473 | 1.004 | 72.29 | 2.718 | 187.9 | 21045 |
| 0.1452 | 447.7 | 6.282 | 1.004 | 72.12 | 2.716 | 184.2 | 18679 |
| 0.1456 | 422.6 | 6.110 | 1.004 | 71.95 | 2.713 | 180.3 | 16747 |
| 0.1461 | 400.3 | 5.954 | 1.003 | 71.79 | 2.710 | 176.4 | 15144 |
| 0.1465 | 380.4 | 5.811 | 1.003 | 71.64 | 2.708 | 172.5 | 13793 |
| 0.1469 | 362.4 | 5.680 | 1.003 | 71.50 | 2.705 | 168.7 | 12643 |
| 0.1473 | 346.2 | 5.559 | 1.003 | 71.36 | 2.703 | 164.9 | 11652 |
| 0.1482 | 311.7 | 5.293 | 1.003 | 71.04 | 2.698 | 156.0 | 9694 |

Table 2, part 5 (cont.)

| $p$ | $t$ | $\lambda'$ | $\lambda''$ | $\eta'$ | $\eta''$ | $\nu'$ | $\nu''$ |
|------|------|------|------|------|------|------|------|
| MPa | °C | mW/Km | | $\mu$Pa s | | $10^{-6}$ m$^2$/s | |
| 0.00500 | 32.88 | 620.0 | 19.08 | 751.1 | 10.09 | 0.755 | 284.6 |
| 0.00550 | 34.59 | 622.6 | 19.21 | 725.6 | 10.14 | 0.730 | 261.4 |
| 0.00600 | 36.17 | 625.0 | 19.32 | 703.2 | 10.19 | 0.708 | 241.9 |
| 0.00650 | 37.63 | 627.2 | 19.42 | 683.3 | 10.24 | 0.688 | 225.3 |
| 0.00700 | 39.01 | 629.1 | 19.52 | 665.6 | 10.28 | 0.671 | 211.0 |
| 0.00750 | 40.30 | 631.0 | 19.62 | 649.6 | 10.32 | 0.655 | 198.5 |
| 0.00800 | 41.52 | 632.7 | 19.71 | 635.0 | 10.35 | 0.640 | 187.4 |
| 0.00850 | 42.67 | 634.2 | 19.80 | 621.7 | 10.39 | 0.627 | 177.6 |
| 0.00900 | 43.77 | 635.7 | 19.88 | 609.5 | 10.42 | 0.615 | 168.9 |
| 0.00950 | 44.82 | 637.1 | 19.96 | 598.2 | 10.45 | 0.604 | 161.0 |
| 0.01000 | 45.82 | 638.4 | 20.04 | 587.8 | 10.49 | 0.594 | 153.9 |
| 0.01200 | 49.43 | 642.8 | 20.32 | 552.3 | 10.60 | 0.559 | 131.0 |
| 0.01400 | 52.56 | 646.5 | 20.57 | 524.4 | 10.70 | 0.531 | 114.4 |
| 0.01600 | 55.33 | 649.5 | 20.79 | 501.6 | 10.78 | 0.509 | 101.7 |
| 0.01800 | 57.81 | 652.1 | 21.00 | 482.4 | 10.86 | 0.490 | 91.74 |
| 0.02000 | 60.07 | 654.4 | 21.19 | 466.1 | 10.94 | 0.474 | 83.66 |
| 0.02200 | 62.15 | 656.4 | 21.37 | 451.8 | 11.00 | 0.460 | 76.97 |
| 0.02400 | 64.07 | 658.1 | 21.54 | 439.3 | 11.07 | 0.448 | 71.34 |
| 0.02600 | 65.86 | 659.7 | 21.69 | 428.1 | 11.12 | 0.437 | 66.53 |
| 0.02800 | 67.54 | 661.1 | 21.84 | 418.1 | 11.18 | 0.427 | 62.37 |
| 0.03000 | 69.11 | 662.4 | 21.99 | 409.0 | 11.23 | 0.418 | 58.73 |
| 0.03500 | 72.70 | 665.1 | 22.31 | 389.6 | 11.35 | 0.399 | 51.37 |
| 0.04000 | 75.88 | 667.3 | 22.61 | 373.6 | 11.45 | 0.383 | 45.75 |
| 0.04500 | 78.74 | 669.2 | 22.89 | 360.2 | 11.55 | 0.370 | 41.31 |
| 0.05000 | 81.34 | 670.8 | 23.14 | 348.6 | 11.64 | 0.359 | 37.71 |
| 0.05500 | 83.73 | 672.1 | 23.38 | 338.6 | 11.72 | 0.349 | 34.73 |
| 0.06000 | 85.95 | 673.3 | 23.60 | 329.7 | 11.79 | 0.341 | 32.22 |
| 0.06500 | 88.02 | 674.3 | 23.81 | 321.8 | 11.86 | 0.333 | 30.07 |
| 0.07000 | 89.96 | 675.2 | 24.01 | 314.7 | 11.93 | 0.326 | 28.21 |
| 0.07500 | 91.78 | 676.0 | 24.20 | 308.2 | 11.99 | 0.320 | 26.58 |

| $a'$ | $a''$ | $Pr'$ | $Pr''$ | $\sigma$ | $b$ | $\beta_v'$ | $\beta_s'$ |
|------|-------|-------|--------|----------|-----|------------|------------|
| $10^{-6}$ m$^2$/s | | — | | mN/m | mm | 1/K | |
| 0.1490 | 283.9 | 5.068 | 1.002 | 70.74 | 2.693 | 147.9 | 8252 |
| 0.1497 | 260.8 | 4.875 | 1.002 | 70.47 | 2.689 | 140.4 | 7151 |
| 0.1504 | 241.4 | 4.706 | 1.002 | 70.22 | 2.685 | 133.7 | 6286 |
| 0.1510 | 224.9 | 4.557 | 1.002 | 69.98 | 2.681 | 127.6 | 5591 |
| 0.1515 | 210.6 | 4.425 | 1.002 | 69.76 | 2.677 | 122.0 | 5021 |
| 0.1521 | 198.1 | 4.306 | 1.002 | 69.55 | 2.674 | 116.9 | 4547 |
| 0.1526 | 187.1 | 4.198 | 1.001 | 69.35 | 2.671 | 112.2 | 4147 |
| 0.1530 | 177.4 | 4.100 | 1.001 | 69.16 | 2.668 | 107.9 | 3806 |
| 0.1534 | 168.7 | 4.010 | 1.001 | 68.98 | 2.665 | 104.0 | 3511 |
| 0.1538 | 160.8 | 3.927 | 1.001 | 68.81 | 2.662 | 100.3 | 3255 |
| 0.1542 | 153.7 | 3.850 | 1.001 | 68.65 | 2.659 | 96.88 | 3030 |
| 0.1556 | 130.9 | 3.593 | 1.001 | 68.04 | 2.650 | 85.41 | 2355 |
| 0.1567 | 114.3 | 3.392 | 1.000 | 67.52 | 2.641 | 76.49 | 1908 |
| 0.1576 | 101.7 | 3.229 | 1.000 | 67.05 | 2.634 | 69.35 | 1592 |
| 0.1584 | 91.75 | 3.094 | 1.000 | 66.62 | 2.627 | 63.50 | 1359 |
| 0.1591 | 83.68 | 2.979 | 1.000 | 66.23 | 2.621 | 58.61 | 1181 |
| 0.1598 | 77.00 | 2.880 | 1.000 | 65.87 | 2.615 | 54.45 | 1040 |
| 0.1603 | 71.37 | 2.793 | 1.000 | 65.53 | 2.610 | 50.88 | 926.8 |
| 0.1608 | 66.57 | 2.716 | 0.999 | 65.22 | 2.605 | 47.77 | 833.8 |
| 0.1613 | 62.41 | 2.647 | 0.999 | 64.92 | 2.601 | 45.03 | 756.2 |
| 0.1617 | 58.78 | 2.585 | 0.999 | 64.64 | 2.596 | 42.61 | 690.6 |
| 0.1627 | 51.41 | 2.453 | 0.999 | 64.00 | 2.586 | 37.59 | 564.4 |
| 0.1634 | 45.79 | 2.346 | 0.999 | 63.43 | 2.577 | 33.68 | 474.1 |
| 0.1641 | 41.36 | 2.257 | 0.999 | 62.91 | 2.569 | 30.52 | 406.8 |
| 0.1647 | 37.76 | 2.181 | 0.999 | 62.43 | 2.561 | 27.93 | 354.8 |
| 0.1652 | 34.77 | 2.115 | 0.999 | 61.99 | 2.554 | 25.76 | 313.6 |
| 0.1656 | 32.26 | 2.057 | 0.999 | 61.58 | 2.547 | 23.90 | 280.2 |
| 0.1660 | 30.11 | 2.005 | 0.999 | 61.19 | 2.541 | 22.31 | 252.6 |
| 0.1664 | 28.24 | 1.959 | 0.999 | 60.83 | 2.535 | 20.92 | 229.6 |
| 0.1667 | 26.61 | 1.918 | 0.999 | 60.49 | 2.530 | 19.69 | 210.0 |

Table 2, part 5 (cont.)

| $p$ | $t$ | $\lambda'$ | $\lambda''$ | $\eta'$ | $\eta''$ | $\nu'$ | $\nu''$ |
|---|---|---|---|---|---|---|---|
| MPa | °C | mW/Km | | $\mu$Pa s | | $10^{-6}$ m²/s | |
| 0.08000 | 93.51 | 676.8 | 24.39 | 302.3 | 12.05 | 0.314 | 25.15 |
| 0.09000 | 96.71 | 678.0 | 24.73 | 291.9 | 12.16 | 0.304 | 22.73 |
| 0.10000 | 99.63 | 679.0 | 25.05 | 283.0 | 12.26 | 0.295 | 20.77 |
| 0.1200 | 104.81 | 680.5 | 25.64 | 268.2 | 12.43 | 0.281 | 17.76 |
| 0.1400 | 109.32 | 681.6 | 26.16 | 256.5 | 12.59 | 0.270 | 15.57 |
| 0.1600 | 113.33 | 682.3 | 26.64 | 246.8 | 12.73 | 0.260 | 13.89 |
| 0.1800 | 116.94 | 682.9 | 27.08 | 238.7 | 12.85 | 0.252 | 12.56 |
| 0.2000 | 120.24 | 683.2 | 27.49 | 231.7 | 12.96 | 0.246 | 11.48 |
| 0.2500 | 127.44 | 683.7 | 28.42 | 217.6 | 13.21 | 0.232 | 9.497 |
| 0.3000 | 133.56 | 683.7 | 29.24 | 206.9 | 13.42 | 0.222 | 8.133 |
| 0.3500 | 138.89 | 683.4 | 29.98 | 198.3 | 13.61 | 0.214 | 7.134 |
| 0.4000 | 143.64 | 683.0 | 30.66 | 191.3 | 13.77 | 0.207 | 6.369 |
| 0.4500 | 147.94 | 682.4 | 31.28 | 185.3 | 13.92 | 0.202 | 5.762 |
| 0.5000 | 151.87 | 681.7 | 31.87 | 180.1 | 14.06 | 0.197 | 5.269 |
| 0.6000 | 158.86 | 680.3 | 32.95 | 171.6 | 14.30 | 0.189 | 4.512 |
| 0.7000 | 164.98 | 678.6 | 33.92 | 164.8 | 14.51 | 0.183 | 3.958 |
| 0.8000 | 170.44 | 676.9 | 34.81 | 159.2 | 14.70 | 0.177 | 3.532 |
| 0.9000 | 175.39 | 675.2 | 35.65 | 154.4 | 14.87 | 0.173 | 3.195 |
| 1.0000 | 179.92 | 673.4 | 36.43 | 150.3 | 15.02 | 0.169 | 2.920 |
| 1.1000 | 184.10 | 671.6 | 37.16 | 146.7 | 15.17 | 0.166 | 2.691 |
| 1.2000 | 188.00 | 669.8 | 37.86 | 143.4 | 15.30 | 0.163 | 2.498 |
| 1.3000 | 191.64 | 668.0 | 38.53 | 140.6 | 15.43 | 0.161 | 2.332 |
| 1.4000 | 195.08 | 666.2 | 39.17 | 138.0 | 15.54 | 0.158 | 2.188 |
| 1.5000 | 198.33 | 664.4 | 39.78 | 135.6 | 15.66 | 0.156 | 2.062 |
| 1.6000 | 201.41 | 662.6 | 40.37 | 133.4 | 15.76 | 0.155 | 1.951 |
| 1.7000 | 204.35 | 660.8 | 40.95 | 131.4 | 15.86 | 0.153 | 1.851 |
| 1.8000 | 207.15 | 659.0 | 41.50 | 129.5 | 15.96 | 0.151 | 1.762 |
| 1.9000 | 209.84 | 657.2 | 42.04 | 127.8 | 16.05 | 0.150 | 1.681 |
| 2.0000 | 212.42 | 655.4 | 42.57 | 126.1 | 16.14 | 0.148 | 1.608 |
| 2.2000 | 217.29 | 651.8 | 43.58 | 123.2 | 16.31 | 0.146 | 1.480 |

| $a'$ | $a''$ | $Pr'$ | $Pr''$ | $\sigma$ | $b$ | $\beta_v'$ | $\beta_s'$ |
|------|-------|-------|--------|----------|-----|-----------|-----------|
| $10^{-6}$ m²/s | | — | | mN/m | mm | 1/K | |
| 0.1670 | 25.18 | 1.880 | 0.999 | 60.16 | 2.525 | 18.61 | 193.2 |
| 0.1675 | 22.75 | 1.814 | 0.999 | 59.55 | 2.515 | 16.76 | 166.0 |
| 0.1680 | 20.78 | 1.757 | 0.999 | 58.99 | 2.506 | 15.26 | 144.9 |
| 0.1687 | 17.77 | 1.665 | 1.000 | 57.99 | 2.489 | 12.94 | 114.6 |
| 0.1693 | 15.56 | 1.592 | 1.000 | 57.10 | 2.475 | 11.24 | 93.96 |
| 0.1698 | 13.87 | 1.533 | 1.001 | 56.31 | 2.462 | 9.936 | 79.10 |
| 0.1702 | 12.54 | 1.483 | 1.002 | 55.59 | 2.450 | 8.903 | 67.94 |
| 0.1705 | 11.45 | 1.441 | 1.003 | 54.93 | 2.439 | 8.064 | 59.29 |
| 0.1712 | 9.448 | 1.357 | 1.005 | 53.46 | 2.414 | 6.523 | 44.40 |
| 0.1716 | 8.070 | 1.294 | 1.008 | 52.21 | 2.392 | 5.471 | 35.03 |
| 0.1720 | 7.060 | 1.244 | 1.010 | 51.10 | 2.373 | 4.707 | 28.64 |
| 0.1722 | 6.284 | 1.203 | 1.013 | 50.10 | 2.355 | 4.126 | 24.05 |
| 0.1724 | 5.669 | 1.169 | 1.016 | 49.19 | 2.339 | 3.670 | 20.59 |
| 0.1725 | 5.167 | 1.141 | 1.020 | 48.35 | 2.324 | 3.302 | 17.92 |
| 0.1727 | 4.397 | 1.094 | 1.026 | 46.84 | 2.297 | 2.745 | 14.07 |
| 0.1727 | 3.831 | 1.057 | 1.033 | 45.51 | 2.272 | 2.343 | 11.45 |
| 0.1727 | 3.396 | 1.028 | 1.040 | 44.32 | 2.250 | 2.040 | 9.566 |
| 0.1726 | 3.051 | 1.003 | 1.047 | 43.22 | 2.229 | 1.803 | 8.155 |
| 0.1724 | 2.769 | 0.983 | 1.054 | 42.22 | 2.209 | 1.613 | 7.065 |
| 0.1722 | 2.535 | 0.965 | 1.062 | 41.28 | 2.191 | 1.457 | 6.199 |
| 0.1719 | 2.336 | 0.950 | 1.069 | 40.41 | 2.173 | 1.327 | 5.499 |
| 0.1717 | 2.166 | 0.936 | 1.077 | 39.58 | 2.157 | 1.217 | 4.921 |
| 0.1714 | 2.018 | 0.925 | 1.084 | 38.80 | 2.141 | 1.122 | 4.438 |
| 0.1711 | 1.889 | 0.914 | 1.092 | 38.07 | 2.126 | 1.040 | 4.029 |
| 0.1707 | 1.774 | 0.905 | 1.100 | 37.36 | 2.111 | 0.969 | 3.679 |
| 0.1704 | 1.672 | 0.897 | 1.107 | 36.69 | 2.097 | 0.906 | 3.376 |
| 0.1700 | 1.580 | 0.890 | 1.115 | 36.05 | 2.083 | 0.850 | 3.112 |
| 0.1696 | 1.497 | 0.883 | 1.123 | 35.43 | 2.070 | 0.800 | 2.880 |
| 0.1693 | 1.422 | 0.877 | 1.131 | 34.83 | 2.057 | 0.754 | 2.676 |
| 0.1685 | 1.291 | 0.867 | 1.146 | 33.71 | 2.032 | 0.677 | 2.331 |

Table 2, part 5 (cont.)

| $p$ | $t$ | $\lambda'$ | $\lambda''$ | $\eta'$ | $\eta''$ | $\nu'$ | $\nu''$ |
|---|---|---|---|---|---|---|---|
| MPa | °C | mW/Km | | $\mu$Pa s | | $10^{-6}$ m²/s | |
| 2.4000 | 221.83 | 648.3 | 44.55 | 120.5 | 16.47 | 0.144 | 1.371 |
| 2.6000 | 226.08 | 644.8 | 45.48 | 118.1 | 16.62 | 0.142 | 1.278 |
| 2.8000 | 230.10 | 641.4 | 46.37 | 116.0 | 16.77 | 0.140 | 1.198 |
| 3.0000 | 233.89 | 637.9 | 47.25 | 114.0 | 16.90 | 0.139 | 1.127 |
| 3.2000 | 237.50 | 634.5 | 48.10 | 112.2 | 17.03 | 0.137 | 1.064 |
| 3.4000 | 240.93 | 631.1 | 48.93 | 110.5 | 17.16 | 0.136 | 1.008 |
| 3.6000 | 244.22 | 627.7 | 49.74 | 108.9 | 17.28 | 0.135 | 0.958 |
| 3.8000 | 247.37 | 624.3 | 50.54 | 107.4 | 17.39 | 0.134 | 0.913 |
| 4.0000 | 250.39 | 620.9 | 51.33 | 106.0 | 17.51 | 0.133 | 0.871 |
| 4.5000 | 257.47 | 612.6 | 53.26 | 102.8 | 17.78 | 0.131 | 0.783 |
| 5.0000 | 263.98 | 604.3 | 55.16 | 100.1 | 18.03 | 0.129 | 0.711 |
| 6.0000 | 275.62 | 588.0 | 58.94 | 95.3 | 18.51 | 0.126 | 0.600 |
| 7.0000 | 285.86 | 572.1 | 62.82 | 91.3 | 18.96 | 0.123 | 0.519 |
| 8.0000 | 295.04 | 556.5 | 66.93 | 87.8 | 19.40 | 0.122 | 0.456 |
| 9.0000 | 303.38 | 541.5 | 71.39 | 84.7 | 19.83 | 0.120 | 0.406 |
| 10.0000 | 311.03 | 527.0 | 76.33 | 81.8 | 20.27 | 0.119 | 0.365 |
| 11.000 | 318.11 | 513.1 | 81.90 | 79.2 | 20.72 | 0.118 | 0.331 |
| 12.000 | 324.71 | 499.9 | 88.27 | 76.7 | 21.18 | 0.117 | 0.302 |
| 13.000 | 330.89 | 487.4 | 95.65 | 74.2 | 21.68 | 0.116 | 0.277 |
| 14.000 | 336.70 | 475.4 | 104.3 | 71.8 | 22.21 | 0.116 | 0.255 |
| 15.000 | 342.19 | 464.0 | 114.6 | 69.5 | 22.79 | 0.115 | 0.236 |
| 17.500 | 354.72 | 437.7 | 151.6 | 63.5 | 24.59 | 0.114 | 0.195 |
| 20.000 | 365.80 | 419.4 | 225.5 | 56.3 | 27.47 | 0.115 | 0.161 |
| 21.000 | 369.88 | 427.2 | 296.0 | 52.4 | 29.49 | 0.115 | 0.148 |
| 21.200 | 370.67 | 434.1 | 321.6 | 51.4 | 30.06 | 0.116 | 0.145 |
| 21.400 | 371.46 | 446.5 | 357.7 | 50.3 | 30.74 | 0.116 | 0.142 |
| 21.600 | 372.24 | 471.4 | 415.2 | 48.9 | 31.59 | 0.116 | 0.139 |
| 21.800 | 373.00 | 544.8 | 533.1 | 48.0 | 33.14 | 0.119 | 0.137 |
| 22.000 | 373.8 | 907.1 | 1408.4 | 47.0 | 38.18 | 0.127 | 0.139 |
| 22.055 | 374.0 | 1419 | | 43.2 | | 0.134 | |

| $a'$ | $a''$ | $Pr'$ | $Pr''$ | $\sigma$ | $b$ | $\beta_v'$ | $\beta_s'$ |
|---|---|---|---|---|---|---|---|
| $10^{-6}$ m²/s | | — | | mN/m | mm | 1/K | |
| 0.1676 | 1.180 | 0.858 | 1.162 | 32.65 | 2.008 | 0.612 | 2.052 |
| 0.1667 | 1.085 | 0.851 | 1.178 | 31.66 | 1.985 | 0.558 | 1.824 |
| 0.1658 | 1.003 | 0.846 | 1.194 | 30.72 | 1.963 | 0.511 | 1.633 |
| 0.1649 | 0.932 | 0.841 | 1.210 | 29.84 | 1.942 | 0.471 | 1.472 |
| 0.1640 | 0.868 | 0.837 | 1.226 | 28.99 | 1.921 | 0.436 | 1.335 |
| 0.1630 | 0.812 | 0.834 | 1.242 | 28.18 | 1.901 | 0.405 | 1.217 |
| 0.1620 | 0.762 | 0.832 | 1.258 | 27.41 | 1.882 | 0.378 | 1.114 |
| 0.1610 | 0.716 | 0.831 | 1.274 | 26.67 | 1.863 | 0.354 | 1.024 |
| 0.1600 | 0.675 | 0.830 | 1.290 | 25.96 | 1.844 | 0.332 | 0.945 |
| 0.1574 | 0.588 | 0.830 | 1.332 | 24.29 | 1.799 | 0.286 | 0.783 |
| 0.1547 | 0.518 | 0.832 | 1.373 | 22.76 | 1.757 | 0.250 | 0.660 |
| 0.1491 | 0.412 | 0.843 | 1.458 | 20.03 | 1.676 | 0.197 | 0.486 |
| 0.1433 | 0.336 | 0.861 | 1.546 | 17.64 | 1.599 | 0.159 | 0.371 |
| 0.1374 | 0.279 | 0.885 | 1.636 | 15.51 | 1.525 | 0.132 | 0.291 |
| 0.1312 | 0.235 | 0.915 | 1.731 | 13.60 | 1.453 | 0.111 | 0.232 |
| 0.1250 | 0.199 | 0.951 | 1.831 | 11.87 | 1.383 | 0.094 | 0.188 |
| 0.1185 | 0.171 | 0.994 | 1.938 | 10.29 | 1.312 | 0.081 | 0.153 |
| 0.1119 | 0.147 | 1.046 | 2.054 | 8.848 | 1.242 | 0.070 | 0.126 |
| 0.1049 | 0.127 | 1.108 | 2.182 | 7.525 | 1.170 | 0.061 | 0.105 |
| 0.0977 | 0.110 | 1.184 | 2.329 | 6.310 | 1.098 | 0.053 | 0.087 |
| 0.0899 | 0.094 | 1.281 | 2.502 | 5.195 | 1.022 | 0.046 | 0.072 |
| 0.0675 | 0.062 | 1.695 | 3.166 | 2.805 | 0.817 | 0.033 | 0.045 |
| 0.0382 | 0.032 | 3.002 | 4.993 | 0.9688 | 0.555 | 0.023 | 0.027 |
| 0.0230 | 0.019 | 5.006 | 7.631 | 0.4096 | 0.404 | 0.019 | 0.021 |
| 0.0195 | 0.016 | 5.936 | 8.806 | 0.3134 | 0.366 | 0.018 | 0.020 |
| 0.0156 | 0.013 | 7.440 | 10.63 | 0.2237 | 0.323 | 0.017 | 0.018 |
| 0.0112 | 0.010 | 10.36 | 13.87 | 0.1416 | 0.273 | 0.016 | 0.017 |
| 0.0061 | 0.006 | 19.50 | 21.83 | 0.1416 | 0.210 | 0.015 | 0.015 |
| 0.0017 | 0.001 | 72.81 | 109.57 | 0.0112 | 0.109 | 0.013 | 0.013 |
| 0 | | ∞ | | 0 | 0 | 0.01214 | |

117

Table 2, part 6

| $p$ | $t$ | $\alpha_p'$ | $\alpha_p''$ | $\delta_h'$ | $\delta_h''$ | $\delta_T'$ | $\delta_T''$ | $\frac{dr}{dT}$ |
|---|---|---|---|---|---|---|---|---|
| MPa | °C | $10^{-3}$/K | | K/MPa | | $10^{-3}$m³/kg | | kJ/kg K |
| 0.00061 | 0.01 | −0.0804 | 3.672 | −0.2417 | 331.0 | 1.022 | −618.1 | −2.389 |
| 0.00065 | 0.8 | −0.0637 | 3.661 | −0.2410 | 324.7 | 1.018 | −606.6 | −2.383 |
| 0.00070 | 1.9 | −0.0440 | 3.648 | −0.2401 | 317.2 | 1.012 | −592.8 | −2.376 |
| 0.00075 | 2.8 | −0.0262 | 3.636 | −0.2393 | 310.4 | 1.007 | −580.2 | −2.371 |
| 0.00080 | 3.8 | −0.0100 | 3.624 | −0.2385 | 304.2 | 1.003 | −568.7 | −2.368 |
| 0.00085 | 4.6 | 0.0048 | 3.613 | −0.2377 | 298.4 | 0.9987 | −558.1 | −2.365 |
| 0.00090 | 5.4 | 0.0184 | 3.603 | −0.2370 | 293.0 | 0.9949 | −548.2 | −2.362 |
| 0.00095 | 6.2 | 0.0309 | 3.594 | −0.2363 | 288.1 | 0.9914 | −539.1 | −2.360 |
| 0.00100 | 7.0 | 0.0426 | 3.585 | −0.2356 | 283.4 | 0.9882 | −530.5 | −2.359 |
| 0.00110 | 8.4 | 0.0637 | 3.568 | −0.2344 | 275.0 | 0.9822 | −514.9 | −2.357 |
| 0.00120 | 9.7 | 0.0823 | 3.552 | −0.2332 | 267.4 | 0.9770 | −501.1 | −2.356 |
| 0.00130 | 10.9 | 0.0990 | 3.538 | −0.2322 | 260.7 | 0.9723 | −488.7 | −2.355 |
| 0.00140 | 12.0 | 0.1140 | 3.525 | −0.2312 | 254.6 | 0.9680 | −477.4 | −2.355 |
| 0.00150 | 13.0 | 0.1276 | 3.513 | −0.2304 | 249.0 | 0.9641 | −467.2 | −2.355 |
| 0.00160 | 14.0 | 0.1401 | 3.502 | −0.2295 | 243.9 | 0.9605 | −457.8 | −2.355 |
| 0.00170 | 14.9 | 0.1516 | 3.491 | −0.2288 | 239.2 | 0.9572 | −449.2 | −2.356 |
| 0.00180 | 15.8 | 0.1622 | 3.481 | −0.2280 | 234.9 | 0.9541 | −441.2 | −2.356 |
| 0.00190 | 16.7 | 0.1721 | 3.471 | −0.2274 | 230.8 | 0.9512 | −433.7 | −2.357 |
| 0.00200 | 17.5 | 0.1814 | 3.463 | −0.2267 | 227.0 | 0.9485 | −426.8 | −2.358 |
| 0.00220 | 19.0 | 0.1983 | 3.446 | −0.2256 | 220.2 | 0.9436 | −414.1 | −2.359 |
| 0.00240 | 20.4 | 0.2133 | 3.431 | −0.2245 | 214.1 | 0.9392 | −402.9 | −2.361 |
| 0.00260 | 21.7 | 0.2269 | 3.417 | −0.2235 | 208.6 | 0.9352 | −392.9 | −2.362 |
| 0.00280 | 22.9 | 0.2392 | 3.404 | −0.2227 | 203.6 | 0.9315 | −383.8 | −2.364 |
| 0.00300 | 24.1 | 0.2505 | 3.392 | −0.2219 | 199.1 | 0.9281 | −375.5 | −2.365 |
| 0.00320 | 25.2 | 0.2609 | 3.381 | −0.2211 | 195.0 | 0.9250 | −367.9 | −2.367 |
| 0.00340 | 26.2 | 0.2705 | 3.370 | −0.2204 | 191.2 | 0.9221 | −360.9 | −2.368 |
| 0.00360 | 27.2 | 0.2795 | 3.361 | −0.2198 | 187.7 | 0.9193 | −354.5 | −2.369 |
| 0.00380 | 28.1 | 0.2880 | 3.351 | −0.2191 | 184.4 | 0.9168 | −348.5 | −2.371 |
| 0.00400 | 29.0 | 0.2959 | 3.343 | −0.2186 | 181.3 | 0.9143 | −342.9 | −2.372 |
| 0.00450 | 31.0 | 0.3138 | 3.323 | −0.2173 | 174.5 | 0.9088 | −330.3 | −2.375 |

| $\dfrac{dp}{dT}\Big|_{sat}$ | $\dfrac{d^2p}{dT^2}\Big|_{sat}$ | $c'_{sat}$ | $c''_{sat}$ | $c'$ | $c''$ | $\chi'_T$ | $\chi''_T$ |
|---|---|---|---|---|---|---|---|
| kPa/K | Pa/K² | kJ/kg K | | km/s | | 1/GPa | |
| 0.0444 | 2.862 | 4.229 | −7.314 | 1.401 | 0.4093 | 0.5101 | 1635500 |
| 0.0469 | 2.998 | 4.222 | −7.279 | 1.405 | 0.4100 | 0.5067 | 1539200 |
| 0.0501 | 3.172 | 4.215 | −7.237 | 1.411 | 0.4107 | 0.5027 | 1429300 |
| 0.0532 | 3.343 | 4.210 | −7.197 | 1.416 | 0.4114 | 0.4991 | 1334100 |
| 0.0564 | 3.511 | 4.205 | −7.160 | 1.420 | 0.4121 | 0.4959 | 1250700 |
| 0.0595 | 3.677 | 4.202 | −7.126 | 1.424 | 0.4127 | 0.4930 | 1177200 |
| 0.0625 | 3.840 | 4.199 | −7.093 | 1.428 | 0.4133 | 0.4904 | 1111800 |
| 0.0656 | 4.000 | 4.196 | −7.062 | 1.432 | 0.4139 | 0.4880 | 1053300 |
| 0.0686 | 4.158 | 4.194 | −7.032 | 1.435 | 0.4144 | 0.4858 | 1000700 |
| 0.0747 | 4.469 | 4.191 | −6.978 | 1.441 | 0.4154 | 0.4820 | 909750 |
| 0.0806 | 4.771 | 4.189 | −6.928 | 1.446 | 0.4164 | 0.4786 | 833970 |
| 0.0865 | 5.067 | 4.187 | −6.882 | 1.451 | 0.4172 | 0.4757 | 769850 |
| 0.0923 | 5.357 | 4.186 | −6.840 | 1.456 | 0.4180 | 0.4731 | 714890 |
| 0.0981 | 5.641 | 4.185 | −6.800 | 1.460 | 0.4188 | 0.4709 | 667260 |
| 0.1038 | 5.920 | 4.185 | −6.763 | 1.463 | 0.4195 | 0.4688 | 625580 |
| 0.1095 | 6.194 | 4.184 | −6.729 | 1.466 | 0.4201 | 0.4670 | 588810 |
| 0.1151 | 6.464 | 4.184 | −6.696 | 1.470 | 0.4208 | 0.4653 | 556120 |
| 0.1207 | 6.729 | 4.184 | −6.665 | 1.472 | 0.4214 | 0.4638 | 526870 |
| 0.1263 | 6.991 | 4.184 | −6.635 | 1.475 | 0.4219 | 0.4625 | 500540 |
| 0.1373 | 7.503 | 4.183 | −6.581 | 1.480 | 0.4230 | 0.4600 | 455070 |
| 0.1482 | 8.002 | 4.183 | −6.531 | 1.484 | 0.4240 | 0.4580 | 417180 |
| 0.1589 | 8.489 | 4.183 | −6.485 | 1.488 | 0.4249 | 0.4562 | 385110 |
| 0.1695 | 8.965 | 4.183 | −6.443 | 1.492 | 0.4257 | 0.4547 | 357630 |
| 0.1801 | 9.432 | 4.183 | −6.403 | 1.495 | 0.4265 | 0.4533 | 333810 |
| 0.1905 | 9.889 | 4.183 | −6.366 | 1.498 | 0.4273 | 0.4521 | 312970 |
| 0.2008 | 10.34 | 4.183 | −6.332 | 1.500 | 0.4280 | 0.4511 | 294580 |
| 0.2111 | 10.78 | 4.183 | −6.299 | 1.503 | 0.4286 | 0.4501 | 278230 |
| 0.2213 | 11.21 | 4.183 | −6.268 | 1.505 | 0.4293 | 0.4493 | 263600 |
| 0.2314 | 11.64 | 4.183 | −6.239 | 1.507 | 0.4299 | 0.4486 | 250440 |
| 0.2563 | 12.68 | 4.183 | −6.172 | 1.511 | 0.4313 | 0.4470 | 222640 |

Table 2, part 6 (cont.)

| $p$ | $t$ | $\alpha_p'$ | $\alpha_p''$ | $\delta_h'$ | $\delta_h''$ | $\delta_T'$ | $\delta_T''$ | $\frac{dr}{dT}$ |
|---|---|---|---|---|---|---|---|---|
| MPa | °C | $10^{-3}$/K | | K/MPa | | $10^{-3}$ m³/kg | | kJ/kg K |
| 0.00500 | 32.9 | 0.3296 | 3.305 | −0.2161 | 168.6 | 0.9039 | −319.5 | −2.378 |
| 0.00550 | 34.6 | 0.3436 | 3.289 | −0.2150 | 163.4 | 0.8995 | −310.0 | −2.380 |
| 0.00600 | 36.2 | 0.3563 | 3.274 | −0.2141 | 158.8 | 0.8956 | −301.6 | −2.383 |
| 0.00650 | 37.6 | 0.3678 | 3.261 | −0.2132 | 154.7 | 0.8919 | −294.0 | −2.385 |
| 0.00700 | 39.0 | 0.3784 | 3.248 | −0.2124 | 150.9 | 0.8885 | −287.2 | −2.388 |
| 0.00750 | 40.3 | 0.3882 | 3.237 | −0.2117 | 147.5 | 0.8854 | −281.1 | −2.390 |
| 0.00800 | 41.5 | 0.3972 | 3.226 | −0.2110 | 144.4 | 0.8824 | −275.4 | −2.392 |
| 0.00850 | 42.7 | 0.4057 | 3.216 | −0.2103 | 141.6 | 0.8797 | −270.2 | −2.394 |
| 0.00900 | 43.8 | 0.4137 | 3.207 | −0.2097 | 138.9 | 0.8771 | −265.4 | −2.397 |
| 0.00950 | 44.8 | 0.4212 | 3.198 | −0.2091 | 136.5 | 0.8746 | −260.9 | −2.399 |
| 0.01000 | 45.8 | 0.4283 | 3.190 | −0.2086 | 134.2 | 0.8723 | −256.7 | −2.401 |
| 0.01200 | 49.4 | 0.4533 | 3.161 | −0.2066 | 126.4 | 0.8639 | −242.4 | −2.409 |
| 0.01400 | 52.6 | 0.4743 | 3.137 | −0.2049 | 120.1 | 0.8568 | −231.0 | −2.416 |
| 0.01600 | 55.3 | 0.4925 | 3.116 | −0.2034 | 114.9 | 0.8506 | −221.6 | −2.423 |
| 0.01800 | 57.8 | 0.5084 | 3.098 | −0.2020 | 110.5 | 0.8450 | −213.6 | −2.431 |
| 0.02000 | 60.1 | 0.5226 | 3.082 | −0.2008 | 106.7 | 0.8400 | −206.7 | −2.437 |
| 0.02200 | 62.1 | 0.5355 | 3.068 | −0.1997 | 103.4 | 0.8354 | −200.7 | −2.444 |
| 0.02400 | 64.1 | 0.5473 | 3.055 | −0.1987 | 100.4 | 0.8312 | −195.4 | −2.451 |
| 0.02600 | 65.9 | 0.5581 | 3.044 | −0.1977 | 97.82 | 0.8273 | −190.7 | −2.457 |
| 0.02800 | 67.5 | 0.5681 | 3.033 | −0.1968 | 95.44 | 0.8236 | −186.4 | −2.463 |
| 0.03000 | 69.1 | 0.5775 | 3.024 | −0.1959 | 93.28 | 0.8202 | −182.5 | −2.469 |
| 0.03500 | 72.7 | 0.5985 | 3.003 | −0.1939 | 88.64 | 0.8123 | −174.1 | −2.484 |
| 0.04000 | 75.9 | 0.6169 | 2.985 | −0.1922 | 84.80 | 0.8054 | −167.2 | −2.498 |
| 0.04500 | 78.7 | 0.6331 | 2.970 | −0.1906 | 81.56 | 0.7991 | −161.4 | −2.512 |
| 0.05000 | 81.3 | 0.6478 | 2.957 | −0.1891 | 78.76 | 0.7934 | −156.4 | −2.525 |
| 0.05500 | 83.7 | 0.6612 | 2.946 | −0.1877 | 76.32 | 0.7881 | −152.1 | −2.537 |
| 0.06000 | 85.9 | 0.6735 | 2.936 | −0.1865 | 74.15 | 0.7832 | −148.2 | −2.549 |
| 0.06500 | 88.0 | 0.6849 | 2.927 | −0.1853 | 72.22 | 0.7786 | −144.8 | −2.561 |
| 0.07000 | 90.0 | 0.6956 | 2.919 | −0.1842 | 70.47 | 0.7743 | −141.7 | −2.572 |
| 0.07500 | 91.8 | 0.7055 | 2.912 | −0.1831 | 68.88 | 0.7701 | −138.9 | −2.583 |

| $\dfrac{dp}{dT}\Big|_{sat}$ | $\dfrac{d^2p}{dT^2}\Big|_{sat}$ | $c'_{sat}$ | $c''_{sat}$ | $c'$ | $c''$ | $\chi'_T$ | $\chi''_T$ |
|---|---|---|---|---|---|---|---|
| kPa/K | Pa/K² | kJ/kg K | | km/s | | 1/GPa | |
| 0.2808 | 13.69 | 4.183 | −6.112 | 1.515 | 0.4325 | 0.4458 | 200410 |
| 0.3050 | 14.66 | 4.183 | −6.057 | 1.519 | 0.4337 | 0.4448 | 182210 |
| 0.3289 | 15.61 | 4.183 | −6.008 | 1.522 | 0.4347 | 0.4441 | 167050 |
| 0.3525 | 16.53 | 4.183 | −5.962 | 1.524 | 0.4357 | 0.4435 | 154220 |
| 0.3758 | 17.43 | 4.183 | −5.920 | 1.527 | 0.4366 | 0.4431 | 143220 |
| 0.3989 | 18.31 | 4.182 | −5.881 | 1.529 | 0.4375 | 0.4427 | 133690 |
| 0.4217 | 19.17 | 4.182 | −5.844 | 1.531 | 0.4383 | 0.4425 | 125350 |
| 0.4444 | 20.02 | 4.182 | −5.810 | 1.532 | 0.4390 | 0.4423 | 117990 |
| 0.4668 | 20.85 | 4.182 | −5.777 | 1.534 | 0.4397 | 0.4422 | 111450 |
| 0.4890 | 21.66 | 4.182 | −5.747 | 1.535 | 0.4404 | 0.4421 | 105590 |
| 0.5111 | 22.46 | 4.182 | −5.718 | 1.536 | 0.4411 | 0.4421 | 100330 |
| 0.5977 | 25.54 | 4.182 | −5.615 | 1.541 | 0.4434 | 0.4423 | 83641 |
| 0.6821 | 28.45 | 4.182 | −5.528 | 1.544 | 0.4454 | 0.4430 | 71722 |
| 0.7646 | 31.22 | 4.182 | −5.454 | 1.546 | 0.4471 | 0.4438 | 62781 |
| 0.8455 | 33.87 | 4.182 | −5.388 | 1.548 | 0.4487 | 0.4448 | 55827 |
| 0.9249 | 36.42 | 4.183 | −5.329 | 1.549 | 0.4501 | 0.4459 | 50262 |
| 1.003 | 38.89 | 4.183 | −5.277 | 1.550 | 0.4514 | 0.4470 | 45709 |
| 1.080 | 41.27 | 4.184 | −5.229 | 1.551 | 0.4525 | 0.4482 | 41914 |
| 1.156 | 43.58 | 4.185 | −5.185 | 1.552 | 0.4536 | 0.4494 | 38703 |
| 1.231 | 45.83 | 4.185 | −5.144 | 1.552 | 0.4546 | 0.4506 | 35950 |
| 1.305 | 48.03 | 4.186 | −5.106 | 1.553 | 0.4556 | 0.4518 | 33564 |
| 1.486 | 53.28 | 4.188 | −5.023 | 1.553 | 0.4577 | 0.4549 | 28792 |
| 1.664 | 58.27 | 4.190 | −4.951 | 1.553 | 0.4595 | 0.4579 | 25211 |
| 1.837 | 63.02 | 4.193 | −4.887 | 1.553 | 0.4612 | 0.4609 | 22426 |
| 2.007 | 67.57 | 4.195 | −4.831 | 1.552 | 0.4627 | 0.4639 | 20197 |
| 2.174 | 71.95 | 4.197 | −4.781 | 1.552 | 0.4640 | 0.4668 | 18374 |
| 2.338 | 76.18 | 4.199 | −4.736 | 1.551 | 0.4653 | 0.4696 | 16853 |
| 2.500 | 80.26 | 4.202 | −4.694 | 1.550 | 0.4664 | 0.4724 | 15567 |
| 2.659 | 84.23 | 4.204 | −4.655 | 1.549 | 0.4675 | 0.4751 | 14464 |
| 2.816 | 88.08 | 4.206 | −4.620 | 1.548 | 0.4684 | 0.4777 | 13508 |

Table 2, part 6 (cont.)

| $p$ | $t$ | $\alpha_P'$ | $\alpha_P''$ | $\delta_h'$ | $\delta_h''$ | $\delta_T'$ | $\delta_T''$ | $\frac{dr}{dT}$ |
|---|---|---|---|---|---|---|---|---|
| MPa | °C | \multicolumn{2}{c}{$10^{-3}$/K} | \multicolumn{2}{c}{K/MPa} | \multicolumn{2}{c}{$10^{-3}\,m^3$/kg} | kJ/kg K |
| 0.08000 | 93.5 | 0.7150 | 2.905 | −0.1821 | 67.43 | 0.7662 | −136.3 | −2.594 |
| 0.09000 | 96.7 | 0.7323 | 2.894 | −0.1802 | 64.85 | 0.7589 | −131.8 | −2.614 |
| 0.10000 | 99.63 | 0.7481 | 2.885 | −0.1784 | 62.63 | 0.7522 | −127.9 | −2.634 |
| 0.12000 | 104.8 | 0.7759 | 2.871 | −0.1752 | 58.97 | 0.7401 | −121.6 | −2.671 |
| 0.14000 | 109.3 | 0.8002 | 2.861 | −0.1724 | 56.05 | 0.7293 | −116.5 | −2.705 |
| 0.16000 | 113.3 | 0.8217 | 2.854 | −0.1698 | 53.63 | 0.7195 | −112.4 | −2.737 |
| 0.18000 | 116.9 | 0.8411 | 2.849 | −0.1674 | 51.58 | 0.7105 | −108.9 | −2.768 |
| 0.20000 | 120.2 | 0.8589 | 2.846 | −0.1652 | 49.81 | 0.7021 | −105.9 | −2.797 |
| 0.25000 | 127.4 | 0.8982 | 2.844 | −0.1603 | 46.25 | 0.6832 | −99.99 | −2.864 |
| 0.30000 | 133.6 | 0.9320 | 2.846 | −0.1559 | 43.51 | 0.6664 | −95.52 | −2.926 |
| 0.35000 | 138.9 | 0.9620 | 2.853 | −0.1519 | 41.32 | 0.6510 | −91.99 | −2.983 |
| 0.40000 | 143.6 | 0.9892 | 2.862 | −0.1482 | 39.50 | 0.6368 | −89.10 | −3.036 |
| 0.45000 | 147.9 | 1.014 | 2.872 | −0.1448 | 37.94 | 0.6235 | −86.68 | −3.087 |
| 0.50000 | 151.9 | 1.038 | 2.884 | −0.1415 | 36.60 | 0.6108 | −84.62 | −3.136 |
| 0.60000 | 158.9 | 1.080 | 2.911 | −0.1354 | 34.36 | 0.5870 | −81.26 | −3.227 |
| 0.70000 | 165.0 | 1.119 | 2.940 | −0.1297 | 32.55 | 0.5648 | −78.63 | −3.312 |
| 0.80000 | 170.4 | 1.155 | 2.972 | −0.1244 | 31.04 | 0.5437 | −76.49 | −3.393 |
| 0.90000 | 175.4 | 1.188 | 3.004 | −0.1193 | 29.75 | 0.5235 | −74.71 | −3.469 |
| 1.00000 | 179.9 | 1.220 | 3.038 | −0.1145 | 28.63 | 0.5040 | −73.20 | −3.542 |
| 1.1000 | 184.1 | 1.251 | 3.073 | −0.1097 | 27.64 | 0.4849 | −71.91 | −3.613 |
| 1.2000 | 188.0 | 1.280 | 3.109 | −0.1052 | 26.75 | 0.4664 | −70.78 | −3.681 |
| 1.3000 | 191.6 | 1.309 | 3.145 | −0.1007 | 25.95 | 0.4481 | −69.79 | −3.747 |
| 1.4000 | 195.1 | 1.336 | 3.181 | −0.0963 | 25.22 | 0.4301 | −68.92 | −3.811 |
| 1.5000 | 198.3 | 1.363 | 3.218 | −0.0920 | 24.56 | 0.4123 | −68.14 | −3.874 |
| 1.6000 | 201.4 | 1.389 | 3.256 | −0.0878 | 23.94 | 0.3947 | −67.44 | −3.935 |
| 1.7000 | 204.3 | 1.415 | 3.293 | −0.0836 | 23.38 | 0.3772 | −66.81 | −3.995 |
| 1.8000 | 207.2 | 1.441 | 3.332 | −0.0795 | 22.85 | 0.3598 | −66.24 | −4.054 |
| 1.9000 | 209.8 | 1.466 | 3.370 | −0.0754 | 22.35 | 0.3425 | −65.72 | −4.112 |
| 2.0000 | 212.4 | 1.490 | 3.409 | −0.0714 | 21.89 | 0.3252 | −65.25 | −4.169 |
| 2.2000 | 217.3 | 1.539 | 3.488 | −0.0634 | 21.05 | 0.2907 | −64.44 | −4.281 |

| $\dfrac{dp}{dT}\Big|_{sat}$ | $\dfrac{d^2p}{dT^2}\Big|_{sat}$ | $c'_{sat}$ | $c''_{sat}$ | $c'$ | $c''$ | $\chi'_T$ | $\chi''_T$ |
|---|---|---|---|---|---|---|---|
| kPa/K | Pa/K² | kJ/kg K | | km/s | | 1/GPa | |
| 2.972 | 91.83 | 4.208 | −4.587 | 1.547 | 0.4694 | 0.4803 | 12671 |
| 3.277 | 99.06 | 4.212 | −4.527 | 1.544 | 0.4711 | 0.4854 | 11276 |
| 3.577 | 106.0 | 4.216 | −4.475 | 1.542 | 0.4726 | 0.4903 | 10160 |
| 4.159 | 118.9 | 4.223 | −4.385 | 1.537 | 0.4752 | 0.4996 | 8485 |
| 4.722 | 131.0 | 4.229 | −4.311 | 1.532 | 0.4775 | 0.5084 | 7288 |
| 5.269 | 142.4 | 4.235 | −4.248 | 1.527 | 0.4794 | 0.5168 | 6389 |
| 5.803 | 153.2 | 4.241 | −4.194 | 1.523 | 0.4811 | 0.5249 | 5690 |
| 6.326 | 163.4 | 4.247 | −4.147 | 1.518 | 0.4826 | 0.5326 | 5131 |
| 7.587 | 187.2 | 4.260 | −4.050 | 1.507 | 0.4858 | 0.5508 | 4123 |
| 8.796 | 208.8 | 4.271 | −3.975 | 1.497 | 0.4883 | 0.5678 | 3450 |
| 9.964 | 228.9 | 4.282 | −3.914 | 1.487 | 0.4905 | 0.5839 | 2969 |
| 11.10 | 247.6 | 4.292 | −3.864 | 1.477 | 0.4923 | 0.5993 | 2608 |
| 12.20 | 265.2 | 4.301 | −3.822 | 1.468 | 0.4938 | 0.6141 | 2327 |
| 13.27 | 282.0 | 4.310 | −3.786 | 1.460 | 0.4952 | 0.6284 | 2102 |
| 15.35 | 313.1 | 4.327 | −3.728 | 1.444 | 0.4975 | 0.6559 | 1764 |
| 17.35 | 341.7 | 4.344 | −3.684 | 1.429 | 0.4993 | 0.6822 | 1522 |
| 19.29 | 368.4 | 4.359 | −3.649 | 1.415 | 0.5008 | 0.7076 | 1341 |
| 21.18 | 393.4 | 4.374 | −3.622 | 1.402 | 0.5020 | 0.7324 | 1199 |
| 23.01 | 417.1 | 4.389 | −3.601 | 1.389 | 0.5030 | 0.7566 | 1086 |
| 24.80 | 439.5 | 4.403 | −3.583 | 1.378 | 0.5039 | 0.7804 | 993.5 |
| 26.56 | 461.0 | 4.417 | −3.570 | 1.366 | 0.5046 | 0.8040 | 916.2 |
| 28.28 | 481.5 | 4.431 | −3.559 | 1.355 | 0.5052 | 0.8273 | 850.7 |
| 29.96 | 501.2 | 4.444 | −3.551 | 1.345 | 0.5057 | 0.8504 | 794.6 |
| 31.62 | 520.2 | 4.457 | −3.545 | 1.335 | 0.5062 | 0.8733 | 745.9 |
| 33.25 | 538.5 | 4.471 | −3.541 | 1.325 | 0.5065 | 0.8962 | 703.3 |
| 34.86 | 556.3 | 4.484 | −3.539 | 1.315 | 0.5068 | 0.9191 | 665.7 |
| 36.45 | 573.5 | 4.497 | −3.538 | 1.306 | 0.5070 | 0.9419 | 632.3 |
| 38.01 | 590.2 | 4.510 | −3.538 | 1.297 | 0.5072 | 0.9648 | 602.3 |
| 39.55 | 606.5 | 4.522 | −3.539 | 1.288 | 0.5074 | 0.9876 | 575.4 |
| 42.58 | 637.8 | 4.548 | −3.545 | 1.271 | 0.5075 | 1.034 | 528.9 |

123

Table 2, part 6 (cont.)

| $p$ | $t$ | $\alpha_P'$ | $\alpha_P''$ | $\delta_h'$ | $\delta_h''$ | $\delta_T'$ | $\delta_T''$ | $\dfrac{dr}{dT}$ |
|---|---|---|---|---|---|---|---|---|
| MPa | °C | $10^{-3}$/K | | K/MPa | | $10^{-3}$m$^3$/kg | | kJ/kg K |
| 2.4000 | 221.8 | 1.587 | 3.567 | $-0.0555$ | 20.29 | 0.2562 | $-63.75$ | $-4.389$ |
| 2.6000 | 226.1 | 1.634 | 3.649 | $-0.0477$ | 19.61 | 0.2214 | $-63.18$ | $-4.496$ |
| 2.8000 | 230.1 | 1.681 | 3.731 | $-0.0399$ | 18.99 | 0.1864 | $-62.70$ | $-4.600$ |
| 3.0000 | 233.9 | 1.727 | 3.815 | $-0.0321$ | 18.42 | 0.1511 | $-62.29$ | $-4.703$ |
| 3.2000 | 237.5 | 1.774 | 3.900 | $-0.0243$ | 17.90 | 0.1153 | $-61.96$ | $-4.804$ |
| 3.4000 | 240.9 | 1.820 | 3.987 | $-0.0166$ | 17.42 | 0.0790 | $-61.68$ | $-4.904$ |
| 3.6000 | 244.2 | 1.867 | 4.075 | $-0.0088$ | 16.97 | 0.0421 | $-61.46$ | $-5.004$ |
| 3.8000 | 247.4 | 1.914 | 4.165 | $-0.0010$ | 16.55 | 0.0046 | $-61.28$ | $-5.102$ |
| 4.0000 | 250.4 | 1.961 | 4.257 | 0.0069 | 16.16 | $-0.0335$ | $-61.15$ | $-5.200$ |
| 4.5000 | 257.5 | 2.081 | 4.493 | 0.0268 | 15.28 | $-0.1323$ | $-60.97$ | $-5.443$ |
| 5.0000 | 264.0 | 2.204 | 4.741 | 0.0471 | 14.52 | $-0.2364$ | $-61.00$ | $-5.686$ |
| 6.0000 | 275.6 | 2.464 | 5.278 | 0.0893 | 13.25 | $-0.4646$ | $-61.53$ | $-6.174$ |
| 7.0000 | 285.9 | 2.749 | 5.881 | 0.1345 | 12.23 | $-0.7256$ | $-62.62$ | $-6.677$ |
| 8.0000 | 295.0 | 3.068 | 6.565 | 0.1834 | 11.37 | $-1.028$ | $-64.22$ | $-7.203$ |
| 9.0000 | 303.4 | 3.429 | 7.351 | 0.2368 | 10.64 | $-1.385$ | $-66.33$ | $-7.762$ |
| 10.000 | 311.0 | 3.847 | 8.263 | 0.2957 | 10.00 | $-1.811$ | $-68.99$ | $-8.367$ |
| 11.000 | 318.1 | 4.338 | 9.339 | 0.3614 | 9.433 | $-2.329$ | $-72.28$ | $-9.028$ |
| 12.000 | 324.7 | 4.926 | 10.63 | 0.4352 | 8.919 | $-2.968$ | $-76.33$ | $-9.764$ |
| 13.000 | 330.9 | 5.647 | 12.19 | 0.5191 | 8.447 | $-3.776$ | $-81.34$ | $-10.60$ |
| 14.000 | 336.7 | 6.552 | 14.14 | 0.6154 | 8.008 | $-4.822$ | $-87.59$ | $-11.56$ |
| 15.000 | 342.2 | 7.725 | 16.64 | 0.7274 | 7.593 | $-6.220$ | $-95.53$ | $-12.69$ |
| 17.500 | 354.7 | 13.05 | 27.53 | 1.110 | 6.619 | $-12.97$ | $-129.1$ | $-16.89$ |
| 20.000 | 365.8 | 32.07 | 63.01 | 1.774 | 5.626 | $-39.68$ | $-230.6$ | $-27.07$ |
| 21.000 | 369.9 | 66.07 | 123.1 | 2.236 | 5.122 | $-91.29$ | $-392.3$ | $-39.79$ |
| 21.200 | 370.7 | 83.57 | 152.8 | 2.367 | 4.993 | $-118.7$ | $-470.4$ | $-45.10$ |
| 21.400 | 371.5 | 112.1 | 208.9 | 2.520 | 4.834 | $-163.9$ | $-616.4$ | $-53.44$ |
| 21.600 | 372.2 | 173.6 | 318.6 | 2.722 | 4.644 | $-263.4$ | $-895.8$ | $-67.56$ |
| 21.800 | 373.0 | 374.6 | 688.4 | 3.013 | 4.385 | $-597.0$ | $-1813$ | $-103.2$ |
| 22.000 | 373.8 | 2802 | 6866 | 3.485 | 4.013 | $-4893$ | $-16220$ | $-261.1$ |
| 22.055 | 374.0 | $\infty$ | | 3.733 | | $-\infty$ | | $-\infty$ |

| $\dfrac{dp}{dT}\Big|_{sat}$ | $\dfrac{d^2p}{dT^2}\Big|_{sat}$ | $c'_{sat}$ | $c''_{sat}$ | $c'$ | $c''$ | $\chi'_T$ | $\chi''_T$ |
|---|---|---|---|---|---|---|---|
| kPa/K | Pa/K² | kJ/kg K | | km/s | | 1/GPa | |
| 45.55 | 667.6 | 4.573 | −3.553 | 1.255 | 0.5075 | 1.080 | 490.1 |
| 48.45 | 696.2 | 4.598 | −3.565 | 1.239 | 0.5074 | 1.127 | 457.4 |
| 51.30 | 723.6 | 4.623 | −3.578 | 1.224 | 0.5072 | 1.174 | 429.3 |
| 54.09 | 750.1 | 4.648 | −3.594 | 1.210 | 0.5068 | 1.222 | 405.0 |
| 56.84 | 775.7 | 4.674 | −3.612 | 1.196 | 0.5065 | 1.271 | 383.8 |
| 59.55 | 800.5 | 4.699 | −3.632 | 1.182 | 0.5060 | 1.321 | 365.1 |
| 62.22 | 824.6 | 4.724 | −3.652 | 1.169 | 0.5055 | 1.372 | 348.5 |
| 64.85 | 848.0 | 4.749 | −3.675 | 1.156 | 0.5049 | 1.424 | 333.7 |
| 67.45 | 870.9 | 4.774 | −3.699 | 1.143 | 0.5043 | 1.477 | 320.5 |
| 73.81 | 925.8 | 4.838 | −3.763 | 1.113 | 0.5026 | 1.615 | 292.7 |
| 80.00 | 978.1 | 4.903 | −3.835 | 1.084 | 0.5007 | 1.762 | 270.7 |
| 91.96 | 1076 | 5.037 | −3.999 | 1.030 | 0.4963 | 2.087 | 238.5 |
| 103.5 | 1169 | 5.180 | −4.189 | 0.9800 | 0.4914 | 2.466 | 216.7 |
| 114.6 | 1258 | 5.332 | −4.407 | 0.9325 | 0.4860 | 2.912 | 201.6 |
| 125.4 | 1345 | 5.498 | −4.656 | 0.8871 | 0.4803 | 3.445 | 191.2 |
| 136.1 | 1431 | 5.680 | −4.941 | 0.8434 | 0.4743 | 4.091 | 184.4 |
| 146.5 | 1518 | 5.884 | −5.268 | 0.8011 | 0.4680 | 4.888 | 180.6 |
| 156.8 | 1606 | 6.115 | −5.646 | 0.7598 | 0.4613 | 5.890 | 179.5 |
| 167.0 | 1696 | 6.382 | −6.087 | 0.7193 | 0.4544 | 7.174 | 181.1 |
| 177.1 | 1791 | 6.696 | −6.610 | 0.6795 | 0.4471 | 8.864 | 185.6 |
| 187.2 | 1892 | 7.076 | −7.240 | 0.6401 | 0.4394 | 11.16 | 193.6 |
| 212.7 | 2192 | 8.548 | −9.641 | 0.5422 | 0.4178 | 22.62 | 238.8 |
| 239.3 | 2686 | 12.38 | −15.60 | 0.4424 | 0.3905 | 70.22 | 402.1 |
| 251.0 | 3113 | 17.35 | −23.14 | 0.3976 | 0.3749 | 165.3 | 674.1 |
| 253.5 | 3240 | 19.47 | −26.28 | 0.3865 | 0.3709 | 216.6 | 806.0 |
| 256.6 | 3529 | 22.49 | −31.54 | 0.3737 | 0.3656 | 302.8 | 1052 |
| 259.4 | 3851 | 27.95 | −40.13 | 0.3552 | 0.3575 | 495.6 | 1521 |
| 262.7 | 4815 | 41.13 | −62.48 | 0.3253 | 0.3408 | 1157 | 3060 |
| 265.8 | 7464 | 106.3 | −155.0 | 0.2708 | 0.3027 | 9806 | 27601 |
| 267.9 | − | ∞ | −∞ | − | | ∞ | |

**Table 3.** Properties of Water at 0.1 MPa from − 30 °C to the Saturation Temperature. Calculated according to [6] (Table A). Values for $t \leqslant 0$ °C relate to subcooled water

**Tafel 3.** Zustandsgrößen von Wasser bei 1 bar von −30 °C bis zur Sättigungstemperatur. Berechnet nach [6] (Tabelle A). Werte für $t \leqslant 0$ °C gelten für unterkühltes Wasser

Table 3                                                                 Tafel 3

| $t$ | $\varrho$ | $c_p$ | $\lambda$ | $\eta$ | $\nu$ | $a$ | $Pr$ |
|---|---|---|---|---|---|---|---|
| °C | kg/m$^3$ | kJ/ kg K | mW/ K m | $\mu$Pas | $10^{-6}$ m$^2$/s | $10^{-6}$ m$^2$/s | — |
| −30.00 | 983.8 | 4.817 | 495.6 | 8661 | 8.804 | 0.1046 | 84.18 |
| −25.00 | 989.6 | 4.561 | 511.5 | 5962 | 6.025 | 0.1133 | 53.17 |
| −20.00 | 993.6 | 4.418 | 523.1 | 4363 | 4.391 | 0.1192 | 36.85 |
| −15.00 | 996.3 | 4.332 | 532.9 | 3339 | 3.351 | 0.1235 | 27.14 |
| −10.00 | 998.1 | 4.277 | 542.3 | 2645 | 2.650 | 0.1270 | 20.86 |
| −9.00 | 998.4 | 4.269 | 544.2 | 2534 | 2.538 | 0.1277 | 19.88 |
| −8.00 | 998.7 | 4.261 | 546.0 | 2429 | 2.432 | 0.1283 | 18.96 |
| −7.00 | 998.9 | 4.254 | 547.9 | 2331 | 2.334 | 0.1289 | 18.10 |
| −6.00 | 999.1 | 4.248 | 549.7 | 2240 | 2.242 | 0.1295 | 17.31 |
| −5.00 | 999.3 | 4.242 | 551.6 | 2154 | 2.155 | 0.1301 | 16.56 |
| −4.00 | 999.4 | 4.236 | 553.5 | 2073 | 2.074 | 0.1307 | 15.86 |
| −3.00 | 999.6 | 4.231 | 555.4 | 1996 | 1.997 | 0.1313 | 15.21 |
| −2.00 | 999.7 | 4.227 | 557.3 | 1924 | 1.925 | 0.1319 | 14.60 |
| −1.00 | 999.8 | 4.222 | 559.1 | 1856 | 1.857 | 0.1325 | 14.02 |
| 0.00 | 999.8 | 4.218 | 561.0 | 1792 | 1.793 | 0.1330 | 13.48 |
| 1.00 | 999.9 | 4.215 | 562.9 | 1732 | 1.732 | 0.1336 | 12.96 |
| 2.00 | 999.9 | 4.211 | 564.8 | 1674 | 1.674 | 0.1341 | 12.48 |
| 3.00 | 1000.0 | 4.208 | 566.7 | 1620 | 1.620 | 0.1347 | 12.03 |
| 4.00 | 1000.0 | 4.205 | 568.6 | 1568 | 1.568 | 0.1352 | 11.60 |
| 5.00 | 1000.0 | 4.203 | 570.5 | 1519 | 1.519 | 0.1358 | 11.19 |
| 6.00 | 999.9 | 4.200 | 572.4 | 1472 | 1.472 | 0.1363 | 10.80 |
| 7.00 | 999.9 | 4.198 | 574.3 | 1428 | 1.428 | 0.1368 | 10.43 |
| 8.00 | 999.9 | 4.196 | 576.2 | 1385 | 1.385 | 0.1373 | 10.09 |
| 9.00 | 999.8 | 4.194 | 578.1 | 1345 | 1.345 | 0.1379 | 9.76 |
| 10.00 | 999.7 | 4.192 | 580.0 | 1306 | 1.307 | 0.1384 | 9.44 |
| 15.00 | 999.1 | 4.185 | 589.3 | 1138 | 1.139 | 0.1409 | 8.08 |
| 20.00 | 998.2 | 4.181 | 598.4 | 1002 | 1.004 | 0.1434 | 7.00 |
| 25.00 | 997.0 | 4.179 | 607.2 | 890.5 | 0.893 | 0.1457 | 6.13 |
| 30.00 | 995.7 | 4.177 | 615.5 | 797.7 | 0.801 | 0.1480 | 5.41 |
| 35.00 | 994.0 | 4.177 | 623.3 | 719.6 | 0.724 | 0.1501 | 4.82 |
| 40.00 | 992.2 | 4.177 | 630.6 | 653.3 | 0.658 | 0.1521 | 4.33 |
| 45.00 | 990.2 | 4.178 | 637.3 | 596.3 | 0.602 | 0.1540 | 3.91 |
| 50.00 | 988.0 | 4.180 | 643.6 | 547.1 | 0.554 | 0.1558 | 3.55 |
| 55.00 | 985.7 | 4.182 | 649.2 | 504.2 | 0.512 | 0.1575 | 3.25 |
| 60.00 | 983.2 | 4.184 | 654.4 | 466.6 | 0.475 | 0.1591 | 2.98 |
| 65.00 | 980.6 | 4.187 | 659.0 | 433.4 | 0.442 | 0.1605 | 2.75 |
| 70.00 | 977.8 | 4.190 | 663.1 | 404.1 | 0.413 | 0.1619 | 2.55 |
| 75.00 | 974.8 | 4.193 | 666.8 | 377.9 | 0.388 | 0.1631 | 2.38 |
| 80.00 | 971.8 | 4.197 | 670.0 | 354.5 | 0.365 | 0.1643 | 2.22 |
| 85.00 | 968.6 | 4.201 | 672.8 | 333.5 | 0.344 | 0.1653 | 2.08 |
| 90.00 | 965.3 | 4.206 | 675.2 | 314.5 | 0.326 | 0.1663 | 1.96 |
| 95.00 | 961.9 | 4.211 | 677.3 | 297.4 | 0.309 | 0.1672 | 1.85 |
| 99.63 | 958.7 | 4.217 | 678.9 | 283.0 | 0.295 | 0.1680 | 1.76 |

# III. Diagrams

# III. Diagramme

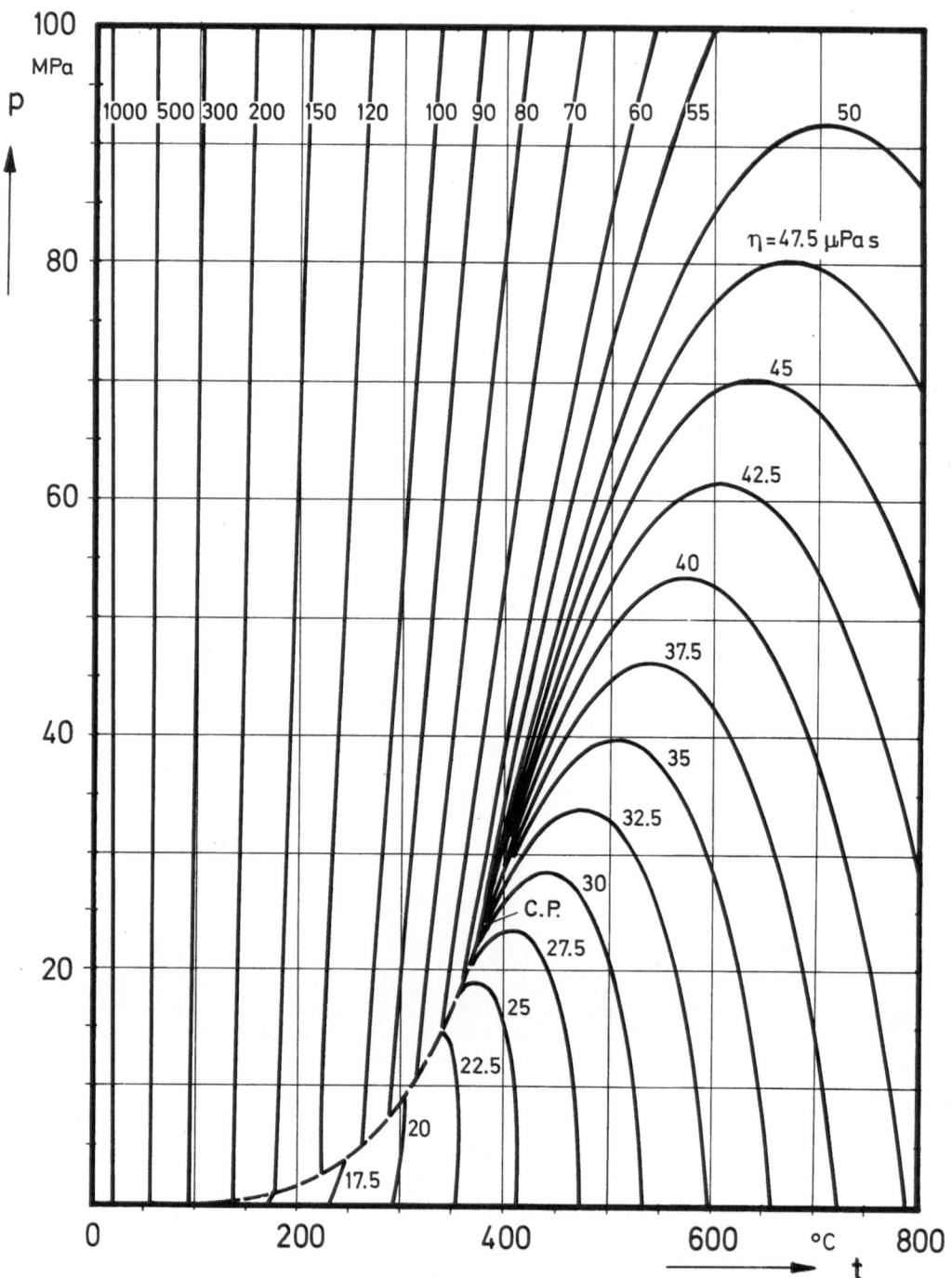

**Fig. 1. Dynamic Viscosity** $\eta$ **in** $\mu$Pa s **in the** $p$, $T$-**Plane.**
**C.P. means Critical Point**

**Bild 1. Dynamische Viskosität** $\eta$ **in** $\mu$Pa s **in der** $p$, $T$-**Ebene.**
**C.P. bedeutet Kritischer Punkt**

130

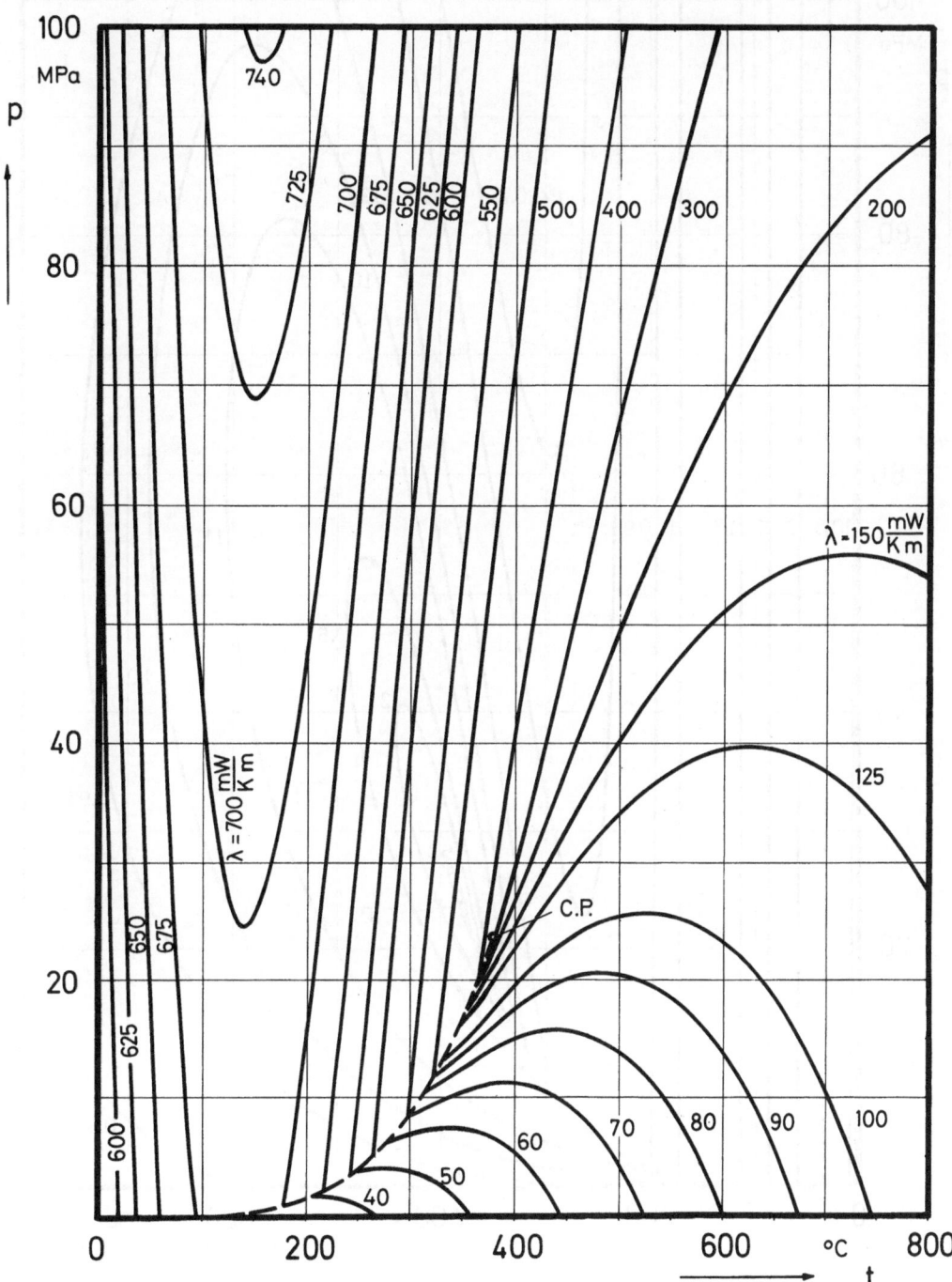

**Fig. 2. Thermal Conductivity λ in mW/Km in the $p$, $T$-Plane.**
**C.P. means Critical Point**

**Bild 2. Wärmeleitfähigkeit λ in mW/Km in der $p$, $T$-Ebene.**
**C.P. bedeutet Kritischer Punkt**

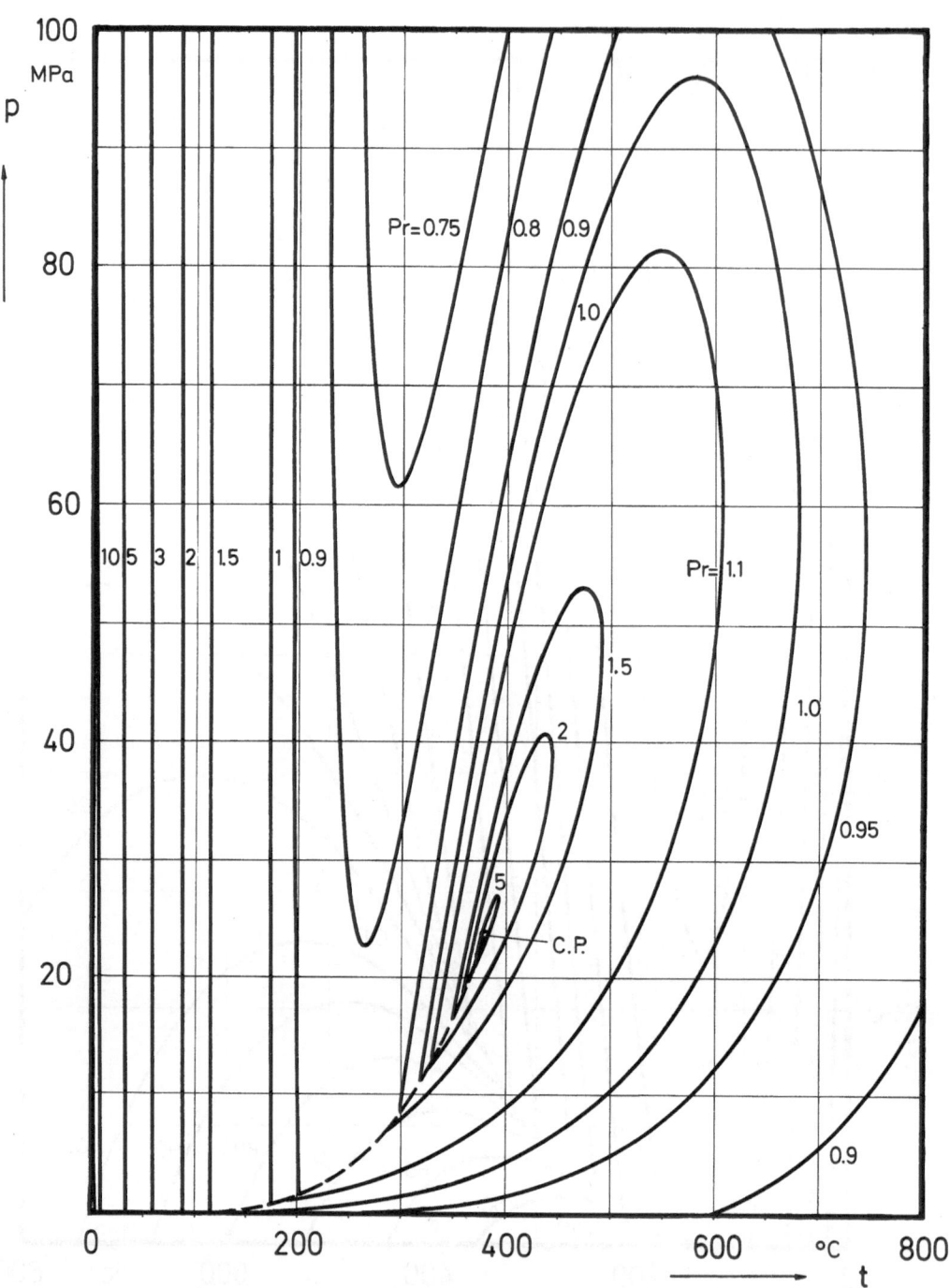

Fig. 3. Prandtl-Number $Pr = \eta\, c_\mathrm{p}/\lambda$ in the $p$, $T$-Plane.
C.P. means Critical Point

Bild 3. Prandtl-Zahl $Pr = \eta\, c_\mathrm{p}/\lambda$ in der $p$, $T$-Ebene.
C.P. bedeutet Kritischer Punkt

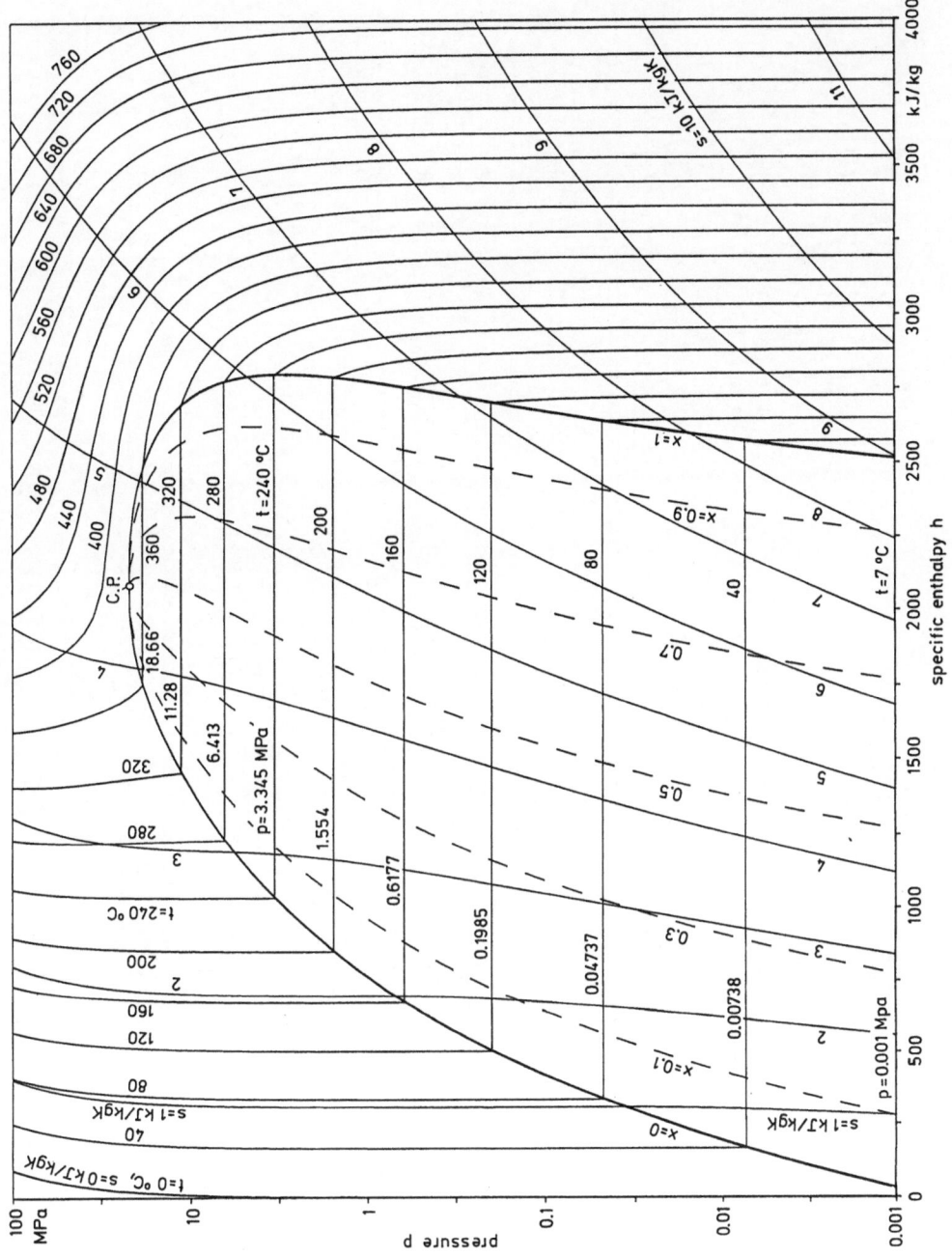

**Fig. 4. Pressure-Enthalpy (log₁₀ $p$, $h$)-Diagram.**
**C.P. means Critical Point**

**Bild 4. Druck-Enthalpie (log₁₀ $p$, $h$)-Diagramm.**
**C.P. bedeutet Kritischer Punkt**

Additional material from *Steam Tables in SI-Units/Wasserdampftafeln,* ISBN 978-3-540-51888-4, is available at http://extras.springer.com